Process Plants

Shutdown and Turnaround Management

T0225514

Process Plants

Shutdown and Turnaround Management

Trinath Sahoo

CRC Press
Taylor & Francis Group
Boca Raton   London   New York

CRC Press is an imprint of the
Taylor & Francis Group, an **informa** business

CRC Press
Taylor & Francis Group
6000 Broken Sound Parkway NW, Suite 300
Boca Raton, FL 33487-2742

First issued in paperback 2017

© 2014 by Taylor & Francis Group, LLC
CRC Press is an imprint of Taylor & Francis Group, an Informa business

No claim to original U.S. Government works

ISBN-13: 978-1-4665-1733-2 (hbk)
ISBN-13: 978-1-138-07750-8 (pbk)

Library of Congress Cataloging-in-Publication Data

Sahoo, Trinath.
    Process plants : shutdown and turnaround management / Trinath Sahoo.
      p. cm.
    Includes bibliographical references and index.
    ISBN 978-1-4665-1733-2 (hardback)
    1. Chemical process control. 2. Chemical plants--Safety measures. 3. Industrial management. 4. Plant shutdowns. I. Title.

TP155.75.S226 2013
338.6'042--dc23
                                        2012037627

Visit the Taylor & Francis Web site at
http://www.taylorandfrancis.com

and the CRC Press Web site at
http://www.crcpress.com

To all the special people who have helped make this book possible:

My parents

Who have taught me to think big

My wife Chinoo

Who makes daily living a delight

My son Sonu

Whose questions make me reflect on many things

My daughter Soha

Who makes life fun.

Contents

Preface

This book is the culmination of the 20 years of experience I have gained in different aspects of shutdowns and turnarounds. It describes the process of managing shutdowns and turnarounds effectively in process plants and will be useful for planners, executors, plant managers, and decision makers.

From the traditional viewpoint, operational shutdowns and turnarounds are maintenance and engineering events. This simplistic view is held by many organizations. A more realistic and holistic perspective, however, recognizes that the impact and scope of shutdown turnaround outages (STOs) extend far beyond the maintenance and engineering functions. STOs can command significant capital and operating budgets. They attract the attention of shareholders and boards of directors and impact inventory supply chains and customer relationships. They are, therefore, whole business events, not simply function-specific ones.

STOs involve both planned and unplanned activities resulting from inspection of a machine that is not accessible or visible during normal operation. The potential for identifying previously unforeseen work requirements discovered during inspection that must be performed within the defined time constraints of the STO adds rapid troubleshooting and decision-making capabilities. The success of the shutdown lies in completing it in the shortest possible time without cost over run and without any safety incident.

This book provides a complete step-by-step guide for managing shutdowns and turnarounds in process plants. It covers all phases of shutdowns and turnarounds, including initiation, planning, executing, controlling, and closing. It also covers many other aspects of shutdowns and turnaround management including performance evaluation and how to make it lean.

Though I have vast experience in the field, I am still learning from managers and practitioners who actually plan and execute shutdown jobs, and any comments on improving the book would be welcome.

Disclaimer

Information contained in this work has been obtained from sources believed to be reliable. However, neither Taylor & Francis Group nor its authors guarantee the accuracy or completeness of any information published herein, and neither shall Taylor & Francis Group nor its authors be responsible for any errors, omissions, or damages arising out of use of this information. This work is published with the understanding that Taylor & Francis Group and its authors are supplying information but are not attempting to render engineering or other professional services. If such services are required, the assistance of an appropriate professional should be sought.

Acknowledgments

I would like to express my gratitude to my parents, who taught me to think big.

I would like to thank my wife, Chinoo, who has managed my home and kids expertly, allowing me to complete the manuscript on time. She has been my friend, confidante, associate, and partner for 15 years. She is interwined in every page and paragraph of this book.

I would also like to thank my son, Sonu, and my daughter, Soha, for their encouragement.

Finally, I would like to thank all the people I have interviewed over the years as well as my associates and friends who have provided valuable feedback.

Author

Mr. Trinath Sahoo is a mechanical engineer of good academic and professional standing. He has 20 years' experience in the technical and management fields.

Mr. Sahoo's key areas of expertise include project engineering design and implementation, process plant maintenance, and troubleshooting and analysis of rotating equipment failure. He has worked as a plant turnaround engineer for various projects and has published papers in various international magazines and conferences.

Mr. Sahoo has thorough knowledge of pump and compressor operation and maintenance; shutdown management, maintenance and reliability; project management and asset management; and advanced maintenance strategies.

He has organized several workshops and has been a speaker in various international conferences in Egypt, the United Kingdom, Canada, Germany, and South Africa.

1

Introduction

The refinery, petrochemical, chemical, and process industries comprise a very capital-intensive sector. Today the process plant equipments have become complex due to the complexity of the technology being employed in building process plants. Machines of all types are now installed in process plants. They are more numerous and more complex compared even to those in the 1960s. Maintenance has to take place at a process plant in order to assure its reliability. It is imperative to take the plant off-line (shutdown) to carry out maintenance of large-scale assets. During this shutdown period, critical inspection, equipment overhaul, repair, and plant modification take place. For a couple of months, a part of or the entire refinery is shut down in order to perform maintenance activities, inspections, and so on. This period will cost the refinery a few million dollars, due to lost production. So it is very important to prepare this shutdown period well.

Process Plant Turnarounds

Refineries' petrochemical and chemical process plants operate round the clock during normal operations; therefore, periodic maintenance is required, along with occasional major overhauls. As in the case of any car, where maintenance and repairs vary significantly, the same is the case with refineries' petrochemical and chemical process plants. For the car owner, major overhauls are highlighted in the new car owner's manual under the preventive maintenance schedule, along with suggested times for more frequent minor maintenance, including fluid changes, belt tightening, mechanical adjustments, and parts or tire replacement. Such maintenance is required to ensure reliable transportation. The same is the case for process plant equipments like compressors, turbines, heat exchangers, and columns. When process plants perform maintenance, they usually need to stop processing raw materials and slow down or stop producing finished products.

The turnaround is a planned, periodic shutdown of a refinery's processing unit (or possibly entire refinery) to perform maintenance, inspection, and repair of equipments that have worn out or broken in order to ensure safe and efficient operations and to replace catalysts that have deteriorated. Often, improvements in equipment or the processing scheme can only be implemented during these turnaround or shutdown periods. Currently, routine turnarounds for refinery units are planned once every 3–5 years.

Activities during a Turnaround

Maintenance activities during a planned turnaround might include

- Routine inspections for corrosion, equipment integrity or wear, deposit formation, integrity of electrical and piping systems
- Special inspections (often arising from anomalies in the prior operating period) of major vessels or rotating equipment or pumps to investigate for abnormal situations
- Installation of replacement equipment for parts of or entire pumps or of instruments that have worn out
- Replacement of catalysts or process materials that have been depleted during operations

Improvement activities could include

- Installation of new, upgraded equipment or technology to improve the refinery processing
- Installation of new, major capital equipment or systems that may significantly alter the refinery process and product output

Outage became more frequent as plants became older and hence accepting shutdown/turnaround became inevitable with so much revenue at stake. This strategy has necessitated changes in plant operating philosophy and inspection technique. Operating companies should use their technical expertise for setting intervals for inspection. So shutdown and turnaround planning and preparation should be carried out more carefully, by assessing plant deterioration and its impact on reliability, planning stocks and logistics, impact of supplies to customers, and checking availability of contractors for plant turnaround activity.

How Shutdown Is Different from Normal Plant Operation

A shutdown is *temporary in nature*, which means that it has a specific start and finish, while operations are ongoing. Operations involve work that is continuous without an ending date and often repeat the same process. There will be *a preferred sequence of execution* for the shutdown tasks (the schedule). The shutdown is a *unique, one-time undertaking*; it will never again be done exactly the same way, by the same people, and within the same environment. This is a noteworthy point, as it suggests that you will rarely have the benefit of a wealth of historical information when you start your project. You'll have to launch your shutdown with limited information or, worse yet, misinformation.

Why Is Shutdown/Turnaround Management So Important?

Shutdown/turnaround costs normally comprise over 30% more of annual maintenance budgets and a delay in start-up can cause a loss of operating profit that exceeds the cost of the shutdown/turnaround. Again, during turnaround, the event requires many people to be diverted and external resources to be brought in.

Shutdowns can be costly in terms of lost production, so a carefully designed plan can reduce costs. Minimizing the duration of the outage can have a major impact on reducing the cost of lost production. Augmenting the resources available to handle shutdown plan scheduling, managing the shutdown activities, and assisting in the start-up of the facilities will minimize the out-of-service time.

Maintenance Philosophy

Machines, are subject to deterioration due to their use and exposure to process and environmental conditions. This deterioration requires to be duly taken care of by various maintenance skills and techniques at minimum cost so that the required use of facilities can be continued and service life extended to the point where maintenance costs become prohibitive and replacement action becomes inevitable. As a result of the above, the expectations from maintenance management have become very high in production management and overall management of the business.

The main objectives of the maintenance management today are

- To keep the plants/equipment, buildings, and service facilities in optimum working conditions
- To keep the downtime of equipment/facility to the minimum so that operating time can be maximized
- To restore equipment/facility to a condition as near to the original condition as possible so that operational use remains within safe limits

- To prolong useful life of equipment/facility while retaining their effective operational use, thus postponing the capital expenditure for their replacement
- To manage maintenance function safely and within optimal costs

Economics of Maintenance Management

The maintenance of equipment/facilities contributes toward optimizing and prolonging their effective use for production activities, thereby enhancing production, and thus sales and profits. Maintenance management also contributes to profitability of the company through reduction of total costs, which comprise

1. Downtime cost/loss of production costs
2. Maintenance material costs
3. Maintenance labor costs

There are four universally recognized maintenance methodologies in use today (see Figures 1.1 and 1.2).

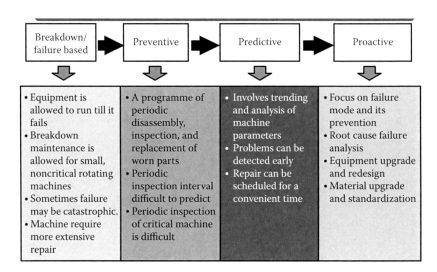

FIGURE 1.1
(See color insert.)

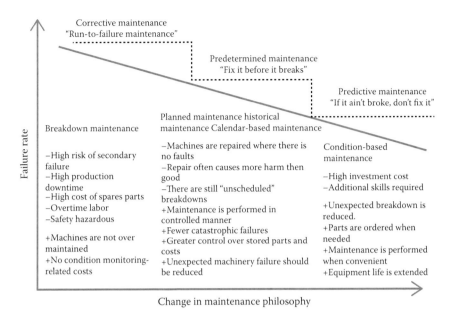

FIGURE 1.2
(See color insert.)

Types of Maintenance Activities

Breakdown maintenance: Repair work is carried out when the equipment has broken down. (See Figure 1.3) Breakdown maintenance is panic or crisis maintenance and is not desired in today's environment. Equipment may not be given proper attention and a breakdown can take place. Such a practice can be recommended only for certain noncritical items whose repair/downtime costs have not been found to be high with this method of operation in previous experiences or where replacement costs permit such a practice. The main disadvantages are

- Uncertain occurrence and repairs
- Uncertain production
- Costly repairs
- Unsafe situation

Preventive maintenance: This is a systematic and planned attention, monitoring, and analysis of the equipment. It is most effective if there are strong age-related failure modes. The guiding principle is "Prevention is better

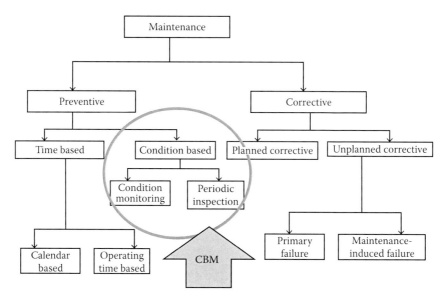

FIGURE 1.3
(See color insert.)

than cure." This consists of various maintenance activities carried out in a deliberately planned manner to prevent or minimize breakdowns to the extent possible. The maintenance activities taken up during preventive maintenance are as follow.

Visual inspection: Visual inspection of equipment is carried out by inspection personnel as well as by maintenance personnel (from their own perspectives) from outside of the equipment. This indicates whether equipment is generally functioning smoothly or whether there are any telltale signs of abnormality like abnormal operating sounds or any leaks of process fluids, cooling liquids, lubricants, etc. From these telltale signs, an experienced maintenance person can infer what is wrong. Depending upon the seriousness of the telltale signs, reparative actions are taken immediately or planned to minimize effects on production schedules.

Lubrication replenishments: Replenishment of lubricating oils to maintain proper quantitative levels or replacing the greases at predetermined intervals in a planned manner provides the required lubrication at sliding or rotating surfaces like in couplings and bearings, thus reducing erosions and extending the operation time of the machinery.

Preventive maintenance plan: Preventive maintenance activities are planned on a regular basis. To ensure smooth operation, the list made out to practice preventive maintenance on these rotary equipments generally incorporates the following items:

- Level and condition of lubricating oil. Replenish or flush the bearing housing and replace the old lubricating oil.
- Renewing grease (for grease-lubricated bearings).
- Cleaning of pump suction strainer.
- Checking the coupling (replenishing/renewing grease).
- Checking rubber pads (renewing if required).

Every week, a certain number of equipments are planned for maintenance, and in this manner all equipments are covered as per the checklist schedule, which is based on the service of the equipments. The weekly report is sent to the rotary section for analysis and records. Such a checklist program contributes to large reductions in premature failures of equipment.

Predictive maintenance: Cost of maintenance can be reduced by deferring maintenance activity until the need has been clearly established. Online condition monitoring using diagnostic instruments/appraisal techniques provides information on the physical health of the equipment and indicates the deterioration trends. A decision can then be taken as to when the equipment can be taken out of service and what quantum of work needs to be done.

Ultrasonic wall thickness measurements: Equipment, storage tanks, pipelines, etc. are subjected to erosion and corrosion, lose their wall thickness, and may fail under operating pressures. Such failures can be prevented through the ultrasonic wall thickness measurement program. Wall thickness of the equipment is measured over critical locations (with the equipment in operation and during outages) periodically in a planned manner. Such data give the trend in the loss of wall thickness which is extrapolated to predict when the wall thickness of the equipment will be reduced to unacceptable levels. Necessary maintenance activities are thus taken to prevent unforeseen failures.

Online vibration measurements: The measurement of vibrations at the driving/ nondriving ends of rotating equipment can be very helpful in predicting the condition of the equipment and its drive. Such analysis predicting the trends in the conditions indicates when equipment needs to be taken out of operation for required maintenance actions to prevent breakdowns.
    Online vibration measurement program:

1. Vibration measurements are taken on a monthly basis on all rotating equipment and on a weekly basis on critical equipment.
2. These readings are recorded on Equipment Vibration History Cards.

3. Equipment is closely monitored when it shows an increase in vibration levels from the normal. In case of an increasing trend, frequency/
   phase analysis of vibrations is carried out, which can indicate

   a. Any increase in the unbalance of rotating parts
   b. Deterioration in the condition of bearings
   c. Increase in misalignment between driving and driven shafts

Testing of lubricating oil/hydraulic oil: Lubricating oil in machinery can get
contaminated with impurities while in use. For example, lubricating oil in
steam turbines can get contaminated with water on account of steam from
the gland finding its way into the bearing housing. The hydraulic system
will thus also get supplied with this water-contaminated oil, accelerating
corrosion and the corrosion products lodging in the closely fitting spaces
between the piston rings and cylinder wall. A consciously implemented
predictive maintenance program of checking the quality of lubricating/
hydraulic oils in rotating equipment to know the trend of buildup of impurities can indicate when the replacement becomes necessary, thus avoiding
breakdowns.

Thermography: Thermography is a technique used to monitor the surface
temperature of equipment. This gives useful inputs about the condition of
the equipment and thus helps in forecasting the time of the outage.

Motor stator analysis and electrical surge testing: This helps in estimating the
condition of the motor stator and insulation.

Motor current signature analysis: This helps in estimating the rotor bars and
the eccentricity.

Process parameters: Process parameters should also be checked as if they are
within the design and at values as per prior experience, then the deterioration of the equipment shall be at expected rates.

Planned maintenance: Equipments undergo erosion/corrosion of their surfaces either by the contact or through the flow of process fluids and even
by the environmental conditions that they are subjected to. There may
also be erosion of surfaces due to mechanical contact during the sliding
and rotating motion of surfaces with respect to each other. The unbalance
of rotating parts can cause vibrations and metal fatigue. Previous inspection of equipment during their overhaul will have given indications of
the rates of erosion/corrosion caused. There are limits within which the
equipment can continue to function safely without losing its mechanical integrity. Knowing the rate of deterioration either through previous
overhauls or through various predictive maintenance practices, equipment can be put on a planned maintenance program, indicating how long
the equipment can be in operating service and when the outage should

be planned for maintenance, inspection, and overhaul. Such outages for planned maintenance become possible either when the equipment is provided with a spare to take over the operating function or during turnaround maintenance of the process unit.

Benefits of planned maintenance: Prevents breakdowns and extends the operating life of equipments:

- By scientifically predicting how long an equipment can be in operating service without a breakdown, the service period can be interrupted in a planned manner and equipment duly rejuvenated to a condition as close to the original as possible for a further useful period of operational service.
- Planned maintenance is based on previous observations and history of equipment. The extent of repair and overhaul work can thus be foreseen pretty well, which enables proper assessment regarding equipment outage time and requirements of various resources, namely, material and manpower.

Turnaround maintenance: Outage of equipment in vital and essential services affects the continuity of operation. Thus, rather than interrupting production every now and then for one or the other equipment and to reduce the overall downtime by clubbing outages of all equipments together, it is planned that all the equipments follow a common cycle of planned maintenance called turnaround maintenance. All equipments can be simultaneously attended to, after which the plant as a whole can operate smoothly till its next turnaround. The main objectives are

- Internal inspection of all static process equipment/piping, which cannot be done during normal operation, ascertaining their condition, and carrying out the required repairs to restore them to a condition as near to the original as possible so that the plant is fully fit and safe to operate till the next turnaround.
- Cleaning, maintenance, repair, and complete overhaul of all rotating/reciprocating equipment, electrical equipment, and instrumentation equipment whose outage can be done only when the plant is not operating.
- Complete overhaul of all pressure-relieving devices (e.g., relief valves on pressure vessels) and resetting to their set pressures.
- Cleaning and complete overhaul of all regulating and controlling instruments whose outage can be done only during a turnaround of the plant.
- Cleaning and overhaul of all electrical switchgears, electric motors, etc., which can be done only during a turnaround.

- Carrying out modification jobs to take care of maintenance and operational needs, upgrading technology through the use of modern equipment, and making the plant more efficient through reducing maintenance and operating costs and increasing the throughput.
- Carrying out items of work that improve the safety aspects of the plant.
- Establishing a database on the condition of equipment/piping and their rates of loss of thickness for future monitoring.

Types of Shutdown

Refinery outages, which derive from a number of situations, may be planned or unplanned. In all cases, part or all of a refinery is taken out of service. For the purpose of this report, four types of outages will be defined: planned turnaround, planned shutdown, unplanned shutdown, and emergency shutdown.

Planned refinery turnarounds are major maintenance or overhaul activities. The frequency of major turnarounds varies by type of unit, but may only need to be done every 3–5 years, for example. Planned turnarounds frequently require 1–2 years of planning and preparation, and sometimes longer when major capital equipment changes are required. The actual turnaround may then last about 20–60 days.

Planned shutdowns are planned, targeted shutdowns of smaller scope than a full turnaround. These mini-turnarounds (or "pit stops"), which help to bridge the gap between planned turnaround intervals, may require 2–6 months of planning and preparation, and the outage may last 5–15 days before returning the processing unit to normal operation.

Unplanned shutdowns are unexpected, but do not require immediate emergency actions. Even well-maintained refinery systems develop unexpected problems. Unplanned shutdowns might result from signs of abnormal or deteriorating process operation. In this situation, the refinery symptom indicates that the affected unit can continue operating for a time, perhaps 3–4 weeks, providing some room for planning, including material and equipment purchases, before the shutdown. Still, this is short notice, and repair plans must be developed on the fly or from previous turnaround procedures or plans. Because of the unexpected nature of such outages, unknown problems may be discovered and cause the unplanned outage to be extended. Unplanned shutdowns are often prolonged due to manufacturing and shipment delays of parts and equipments.

Planned turnarounds and shutdowns can also result in unplanned outage time. Sometimes when a processing unit is brought down for planned maintenance, other problems are discovered that may extend the time offline. When planned turnaround is more difficult than anticipated, this could result in unplanned outage time. Sometimes the unit may have to be brought down several times before it is able to run steadily at full operation. Such problems can sometimes extend over several months.

Emergency shutdowns occur when a unit or entire refinery must be brought down immediately without warning. For example, a fire or power outage could create such a shutdown requirement. A recent survey of FCC units indicated that the biggest reason for unplanned and emergency shutdowns of these units was unexpected loss of utilities (e.g., electricity) to the unit. Unsafe conditions, such as potential severe weather, can also require emergency shutdowns until the weather danger passes, although some weather conditions such as evolving hurricanes allow for more planning. Emergency shutdowns can present some of the largest safety issues, and increase the potential for mechanical damage as a result of the fast shutdown.

In all shutdown cases, when major fuel-producing units are offline, production of gasoline or distillate fuels may be reduced.

Shutdown of Key Refinery Units Affecting Product Pattern

Crude oil feeds into the distillation tower. When crude oil is distilled, before it goes to the various units downstream of the crude oil tower, it is separated into various boiling-range streams. The lighter streams eventually go to gasoline, the middle streams to distillates (with sulfur removal), and the heavier, high-boiling-point streams go to residual fuel or are sent to other units like the hydrocracker, FCC, or coking unit to be broken down to make lighter, higher-valued products.

The units downstream of the distillation tower are where most of the crude oil molecules are transformed (i.e., refined) into higher-quality, higher-valued transportation fuels. For example, the light straight-run gasoline stream that goes directly into the gasoline pool only constitutes about 5% of the total gasoline produced. The remaining gasoline volumes come from the downstream units.

A refinery turnaround does not usually require a complete shutdown of the refinery. However, when significant fuel-producing units are taken offline, gasoline, kerosene, jet fuel, diesel fuel, or heating oil production may be affected.

One product, such as gasoline, may actually be made of different streams from the refinery. For example, gasoline is a mixture of reformate, alkylate, hydrockate gasoline, FCC gasoline, and potentially some straight-run gasoline directly from the distillation tower. Each of these gasoline streams has different emission and driving-performance characteristics. Gasoline for retail is a blend of these streams designed to meet emission requirements and provide good driving performance. When one significant gasoline production unit such as the FCC unit is out of service, the refinery is hampered in its ability to produce gasoline with adequate driving and environmental characteristics, even though the reformer and other gasoline-producing units are still operable. Still, with planned turnarounds, this loss can be addressed to some degree with advance planning.

An outage in any major refinery unit can affect the production of finished products, such as gasoline or distillate. Furthermore, the integration of these units means that the shutdown of one unit for repairs or maintenance can result in the shutdown or reduced operations of other units because of the blending of products or one unit giving feed to another unit.

Strategic Issues

The quest for production without regular plant shutdown goes on, but there is always a need to carry out significant maintenance activity in the form of plant turnarounds. The management have strategic and tactical plans.

The strategic issues regarding plant turnaround are

1. The need for turnaround
2. How to plan, prepare, and execute the turnaround

The process industry always tries to lengthen the interval between turnarounds. Equipment history and current knowledge of the plant equipment asset conditions are major keys for developing the plant turnaround strategy. This strategy has necessitated changes in plant-operating standards and inspections techniques.

To optimize plant run time and avoid major unscheduled outages, a long-term plant turnaround frequency strategy should be developed. The framework will recognize that the plant turnaround procedure has five fundamental phases: strategic planning, detailed planning, organizing, execution, and closeout.

The plant turnaround procedure is a continuous process and the turnaround team position and responsibility needs to overlap from turnaround to turnaround. From turnaround to turnaround, there should be a commitment to a continuous budget and a standardized cost control structure.

The management team should understand that only those procedures requiring plant shutdown should be included in the turnaround that will be statistically or economically beneficial in extending the process run time.

Shutdown durations are kept extremely short compared with other "time-bound" projects because of the need to keep the unit downtime to a minimum. It is vitally important that the shutdown work list be kept as short as possible. Keeping the list short is both a means to reduce cost and the primary method for focusing on work that can only be performed during a major outage. All other work is deferred to a time outside of the shutdown window. Major shutdowns should not be used to avoid periodic minor downtime. Minor downtime is an important part of proactive maintenance because it provides an opportunity to perform periodic preventive maintenance and repairs that cannot be done during operation. Each plant must determine the frequency of its planned minor downtime events, and once an effective cycle is established, it should not be needlessly interrupted.

The compressed time "allowed" by the management generally demands high manning intensity, which in turn dictates unconventional work patterns like overtime and shift work. In addition to these manning challenges, material deliveries are almost invariably critical, with every item having the potential to cause the shutdown completion date to slip. In fact, there is a very strong correlation between the length of time allocated to shutdown planning and the shutdown's ultimate success.

Each plant will need a plant turnaround management process document tailored to its specific need. These needs include the type of plant, the geographical location of the plant, the size of the plant, and the general complexity of the plant outage. What also requires to be established is the frequency of shutdown, a review of the nondestructive examination (NDE) comparison report finding and the regular preventive maintenance inspection programs, and a risk assessment of the process system and consideration of influences outside the plant. Major plant outage frequency should be evolutionary rather than past practices.

Outside influence also has an impact on turnarounds like government and regulatory requirements, time of year (holidays/Christmas), availability of work execution contractor, feedstock, and market condition. Sometimes a shortage of feedstock requiring the plant to run at less than the economic capacity or to shutdown altogether provides an opportunity for a major, short, practical plant shutdown.

Sometimes plant turnaround can become a potential hazard to plant reliability. This may happen if the turnaround is not properly planned, prepared, and executed due to poor decisions, bad workmanship, and use of incorrect materials or because of damage done to the plant during overhaul.

- During normal running of the plant, experienced people usually carry out familiar tasks using well-defined procedures, but during plant shutdown, one could come across hazardous procedures and unfamiliar events. In such situations, the probability of accidents increases.

- "Open an equipment if the probability of fixing a defect is greater than the probability of causing a damage."
- A long-range strategic plan will provide milestone targets for project and maintenance engineering to install future tie-ins, make major repairs and alterations, or replace major equipments.

Taking all these into consideration, the senior management of the company can take a decision on whether to opt for plant turnaround or not.

The shutdown management process has five phases.

Initiation Phase

In this phase, the need is identified. An appropriate response to the need is determined and described. (This is actually where the shutdown begins.) The major deliverables and the participating work groups are identified. The team begins to take shape. Issues of feasibility (*can* we do the shutdown of the unit?) and justification are addressed.

Planning Phase

Here the solution is further developed in as much detail as possible. Intermediate work products (interim deliverables) are identified, along with the strategy for producing them. Formulating this strategy begins with the definition of the required elements of work (tasks) and the optimum sequence for executing them (the schedule). Estimates are made regarding the amount of time and money needed to perform the work and when the work is to be done. Planning has more processes than any of the other shutdown management processes. The executing, controlling, and closing process groups all rely on the planning process group and the documentation produced during the planning processes in order to carry out their functions. Shutdown managers will perform frequent iterations of the planning processes prior to shutdown completion. Therefore, planning must encompass all areas of shutdown management and consider budgets, activity definition, scope planning, schedule development, risk identification, staff acquisition, procurement planning, among other functions.

Executing

During the *execution phase*, the prescribed work is performed under the watchful eye of the shutdown manager; it involves putting the shutdown plans into action. It's here that the shutdown manager will coordinate and direct all resources to meet the objectives of the plan. Progress is continuously monitored and appropriate adjustments are made and recorded as variances from the original plan. Throughout this phase, the shutdown team remains focused on meeting the objectives developed and agreed upon at the outset of the shutdown.

Controlling

The controlling phase is where performance measurements are taken and analyzed to determine if the shutdown is staying true to the project plan. If it's discovered that variances exist, corrective action is taken to get the project activities aligned with the project plan. This might require additional passes through the planning process to realign to the project objectives.

Closing

The closing process is probably the most often skipped in shutdown management. Once the shutdown objectives have been met, most of us are ready to move on to the plant operation. However, closing is important as all the information is gathered at this stage and stored for future reference. The documentation collected during closing processes can be reviewed and utilized to avert potential problems during future shutdowns. Contract closeout occurs here, and formal acceptance takes place.

Turnaround Evaluation

The turnaround evaluation cycle starts with the end of the turnaround and completes with the inspection outlook report and a turnaround report. The inspection outlook report provides details of the jobs carried out and the anticipated job scope for the next turnaround based on the inspection carried out during the turnaround. The turnaround report is a comprehensive document where the lessons learned in the preceding turnaround are put in place as part of the ongoing continuous improvement cycle.

Turnaround Timeline

Turnarounds are large and expensive. A refinery will have a shutdown organization that plans for the event 1–2 years in advance (longer if major capital equipment or process changes are involved). While these organizations vary among companies, in larger refineries, the shutdown organization may be a permanent group, which moves their efforts from one process unit's turnaround (e.g., FCC) in year 2008 to another's turnaround (say crude distillation tower) in 2010. The shutdown team would typically use sophisticated planning and scheduling software (such as Impress, Primavera, ATC Professional, and SAP). These computer programs can coordinate thousands of individual maintenance jobs (which may include 100,000+ separate steps in those jobs), including the steps performed by contractors.

Generic Turnaround Timeline

While the turnaround itself may only last a month or two, planning for the event would typically begin several years ahead. Companies typically schedule major FCC turnarounds every 4 or 5 years. Every second or third turnaround might require more significant work to replace worn equipment.

36–30 Months Prior to Turnaround

A decision is made about whether a major new technology or long lead-time equipment will be part of the turnaround. If so, then scheduling must be determined for equipment orders and special aspects (e.g., special cranes, personnel, and procedures). New vessels, compressors, or turbines can often take 2 or more years to design and fabricate. In addition, some complicated crane construction and associated work, which must be reserved well in advance, can only be done by a few companies. If special work is not required, the turnaround is treated as normal, with main planning starting 2 years in advance.

24 Months Prior to Turnaround

The project/shutdown concept is defined and approved, including equipment upgrades that may be needed and changes to equipment to improve operations and/or yields. Detailed designs for any equipment modifications must be categorized. Some vessel modifications and piping revisions may require an outside engineering firm and extensive interaction with operations. Refineries have a management process, referred to as the "management of change" (MOC) process that will look at safety, environmental, and operation changes that these modifications can cause. In the past decade, this management process has become a serious, lengthy examination of any changes that are proposed. Even information systems for the project could become an issue. For this MOC process to work, the final engineering design and plans for how the change will be operated must be completed at least 6 months before the turnaround begins.

18 Months Prior to Turnaround

The detailed project/shutdown planning process gets under way. Over a 6-month period, equipment inspections may be performed, critical equipment maintained, and systems reviewed to flesh out the specific activities that need to be performed in more detail. Organization of the complex management of the project begins. Long lead-time materials may be purchased, and negotiations begin with the various contractors that deal with refinery turnarounds. In addition, progress is checked on any new vessels, compressors, or turbines that have been ordered. The MOC schedule is also reviewed for (1) any new equipment, (2) equipment or piping modifications, and (3) new operating procedures for equipment after the turnaround.

12 Months Prior to Turnaround

Equipment deliveries and modification designs are checked regularly. The basic scope of the shutdown (maintenance and minor modifications) is finalized, and remaining contracting with outside shop facilities is negotiated. These outside shops work on compressors, turbines, control valves, relief valves, heat exchangers, critical piping systems, etc. Management assures that the MOC process begins on new equipment, equipment and piping modifications, and new operating procedures.

Last Months Prior to Turnaround

The turnaround organization takes control of schedules and management of the activities of the event. Often this organization will be set up on site several months prior to the turnaround and begin to establish the special communication, tracking, and physical aspects necessary for a well-coordinated turnaround.

2

Initiating the Turnaround

Initiation

Initiation is the first process in a turnaround life cycle. You can think of it as the official kickoff, which acknowledges that the shutdown process has begun. The initiation process has several inputs: important tasks, strategic plan, contractor selection criteria, and historical information. Each of these inputs is processed using tools and techniques to produce the final outputs, one of which is the shutdown charter. Initiation culminates in the publication of a shutdown charter. During the initiation period, turnaround parameters are defined, core personnel appointed, and basic data organized. The initiation process lays the groundwork for the planning process group that follows initiation. A high percentage of shutdowns fail due to poor planning or no planning. Properly planning the shutdown work up front dramatically increases the project's chance for success. Since initiation is the foundation of planning, the importance of initiation is self-evident.

Most organizations follow a formal procedure to select and prioritize shutdown time, duration, and major jobs to be carried out during the period. A steering committee is formed who is responsible for review, selection, and prioritization. A steering committee is a group of personnel comprised of senior managers and sometimes midlevel managers who represent each of the functional areas in the organization. The steering group takes responsibility for the long-term strategy for turnarounds. The committee meets at regular intervals to review current performance and formulate high-level strategies for the management of turnaround events and the long-term turnaround program.

Developing a Shutdown Overview

One of the first steps a shutdown manager will take in the initiation process is to develop a shutdown overview. Some organizations call this a concept document. The four inputs described in the previous section will help to outline the shutdown overview and will be used again when formulating the final charter. At the completion of the initiation process, the organization commits to fund the shutdown and provide the necessary resources to carry it out. The stakeholders have the greatest ability to influence the shutdown outcome at this stage because nothing is cast in stone as yet. There is still time for them to negotiate requirements and deliverables.

Strategic Plan

Part of the responsibility of a shutdown manager during the initiation process is to take into consideration the company's strategic plan.

Strategy

During shutdown, there will be draining of resources, there will be potential safety hazard, and there will be unforeseen problems. Taking all of these into consideration, the senior management of the company creates a business strategy for managing turnarounds. Can we eliminate the maintenance turnarounds altogether? Thereafter, if it is proven that the event is absolutely necessary, senior management should ensure that the turnaround is aligned with maintenance objectives, production requirements, and business goals.

A long-term plan for shutdown management should be outlined in the budget forecast 3–10 years before execution. The long-term plan contains fairly detailed lists of the major work that must be performed during each scheduled major shutdown. For instance, boiler inspections, relining of large tile tanks, sewer repairs, and electrical power distribution system inspections are estimated in the long-term plan. Funds must also be included for smaller repairs required during the shutdown, which are often estimated as a lump sum figure.

The long-term plan is the tool for controlling the scope of each outage. Long-term planning is a critical and often overlooked piece of the proactive approach to maintenance. Without a long-term plan, major repairs and inspections often do not get adequate attention until it is too late to properly prepare for their execution.

In addition, an operating budget should be constructed annually. In order to accurately budget for a major shutdown, the scope, duration, and timing of the outage should be supplied before the operating budget is approved. This means that any major shutdown is scoped to an accuracy of ±10% at least 18 months before it is scheduled to take place in order for the budget process to proceed. This is the short-term plan for shutdown management. The short-term plan is developed using the long-term plan as a starting point. In addition to the major repairs, the short-term plan must include detailed lists and estimates for the smaller, less costly repair work. As the budget and shutdown plan enter the approval process, it should be very clear what the upcoming outage will accomplish. It is relatively easy to establish which projects are driving the outage. If this is not the case, then a shutdown is not justified.

Turnaround evaluation: The turnaround evaluation cycle starts with the end of the turnaround and completes with the inspection outlook report and a turnaround report. The inspection outlook report provides details on the jobs carried out and the anticipated job scope for the next turnaround based on the inspection carried out during the turnaround. The turnaround report is a comprehensive document where the lessons learned in the preceding turnaround are put in place as part of the ongoing continuous improvement cycle.

Conducting a Kickoff Meeting

The typical launch of a shutdown begins with a kickoff meeting involving the major players responsible for planning, including the shutdown manager, managers for certain areas of knowledge, subject matter experts (SME), and functional leads. The formal kickoff meeting is intended to recognize the "official" formation of the team and the initiation of shutdown.

The major players are usually authorized by their functional areas to make decisions concerning timing, costs, and resource requirements. Some of the items discussed in the initial kickoff meeting include

- Initial discussion of the scope of the shutdown, including both the technical objective and the business objective
- The definition of success on this shutdown
- The assumptions and constraints as identified in the shutdown charter
- The organizational chart (if known at that time)
- The participants' roles and responsibilities

Plant Turnaround History

An examination of past turnaround performance and of subsequent plant performance will indicate whether the turnarounds have provided the protection expected. If not, the situation must be reassessed to find out why and how a new rationale for turnarounds developed. Past events should also be analyzed to ascertain the ratio of emergent work (which only arises after the execution phase has begun) to planned work, and the extent to which emergent work increased the planned expenditure (the norm is between 5% and 10%).

Current Performance

The scope of the turnaround work may be dictated by the current operating performance of the plant and the level and effectiveness of preventive maintenance carried out in the periods between turnarounds. Regarding the work list that is generated by the current performance of the plant, care should be taken to ensure that the remedial work requested will actually address the problem being experienced.

Establish Shutdown Charter

The shutdown charter defines the vision, objectives, scope, and deliverables for the shutdown. It also provides the organization structure (roles and responsibilities) and a summarized plan of the activities, resources, and funding required to undertake the shutdown. Finally, any risks, issues, planning assumptions, and constraints are listed.

The shutdown is treated like any other project using classic project management techniques. Applying project management techniques for planning, executing, monitoring, and controlling the progress of the shutdown can delineate various scheduling, resources, and cost questions such as

- Is the amount of work doable within the allotted period?
- What are the critical path jobs for completing the shutdown on schedule?
- Have enough resources (personnel, time, money) been allocated?
- How much do I anticipate the shutdown to cost? Can I stay within budget?
- If, on the other hand, it is recognized that all of the tasks on the list can be investigated with a view to eliminating, or at least minimizing, them, then the first step has been taken toward challenging the necessity for turnarounds.

Establish the Shutdown Vision and Objectives

The shutdown must reflect the business goals of the organization. Typical vision statements might read thus: "Complete the December shutdown on time and within budget with no safety incidents."

Within the shutdown plan, objectives and expectations must be established early for the entire operation. Objectives should be concise and measurable as well as applicable to each phase of the shutdown and reflect the outcome established by the vision. Some examples of objectives include

- Limit new or growth work to less than 25% of total shutdown work.
- Eighty-five percent of shutdown work will be determined by inspection and condition monitoring versus historical data.
- Zero safety incidents by contractors or plant work force.

Shutdown Requirement

- Is the shutdown work really necessary and vital to the safety, quality, or production efficiency of the plant?
- Is it desirable at this time in the life of the plant?
- Is it too late or too soon?

Initiating

Successful turnarounds are only achieved when the work scope is well defined, organized, managed, frozen early, and the planning is of a high level.

This enables labor and equipment resources to be well organized and managed.

Planning for a major turnaround virtually begins immediately after the finalization of the last turnaround report and the cycle is as long as the longest lead time for any materials and equipment. This lead time can be as high as 24 months for specific items. These items can be rotors/stators for rotating equipment, cast lift pipe sections for a cracking unit, heater tubes, regenerator cyclones, exchanger parts, and special alloy steel components. It is easy to see that beginning the turnaround planning cycle 6–12 months out can be, and often is, too late. Late procurement effort may result in delaying the turnaround and increase the possibility of an unscheduled shutdown while at the same time incurring costly premium payments. The planning cycle consequently needs to begin as early as possible.

- Initially, one should decide who is to be responsible for the ongoing activities—this should be a senior manager.
- Start by reviewing the last unit turnaround report and identify from inspection reports the material and other items to be ordered.
- Initiate the appropriate procurement activity.
- At the appropriate time call the functions together and solicit items for the preliminary job lists.
- Set milestones to be met and confirm the need to procure additional items identified from the turnaround report.

With this listing and with the input from the strategic planners, decide the best timing slot for the turnaround. Factors impacting here are market supply, unit condition, and material and labor availability. The latter is particularly important when other nearby refineries or plants draw from a common labor pool. At this point, having the base data, the turnaround work period can be estimated and preliminary key performance indicators (KPIs) set.

Scope Statement

At this time, you require a scope statement for the shutdown. The purpose of the scope statement is to document the shutdown goals, deliverables, and requirements so that they can be used as a baseline for future decisions. The work scope is the foundation upon which all other aspects of the turnaround rest and will have a major influence in determining the final cost. If the shutdown charter is written well, it's simply a matter of transferring the goals and deliverables information from the charter to the scope statement. The scope statement is the baseline for the shutdown. This means if questions arise or changes are proposed, they can be compared to what's documented in the scope statement.

In the initiating phase, one must define the boundaries of shutdown:

- What are the goals and objectives?
- Who are the principal participants?
- When must it be finished?
- Why is the shutdown being launched?
- What are the constraints/limitations?
- Coordination requirements
- Levels of support from participants
- Major assumptions
- Major responsibilities
- Milestone dates
- Quality criteria

Assumptions

This section lists any unsubstantiated ideas about the project. Assumptions may, for example, relate to levels of internal support or existing or market conditions. Assumptions are used in planning.

Constraints

Rarely does a project have unlimited resources at its disposal. Money, time, people, equipment, supplies, and facilities are often limited in quantity and quality. Recognizing such limitations early on enables realistic planning.

Appoint Shutdown Team

At this point, the scope of the shutdown has been defined in detail and the shutdown team is ready to be appointed. A shutdown manager can be appointed prior to the establishment of the shutdown team. The shutdown manager documents a detailed job description for each role and appoints a human resource personnel to each role based on his or her relevant skills and experience. Once the team is "fully resourced," the shutdown office is ready to be set up.

Major Turnaround Milestones

Major turnaround milestones are mentioned here in brief comprising all the three cycles of the turnaround. This will help as a ready reckoner of the entire turnaround process. Monitoring these milestones will give complete control of the turnaround activities. The turnaround outlook report should contain details about the jobs accomplished in the turnaround and all the previous outstanding jobs pertaining to the plant. This helps in documenting the pending jobs when the quantum is at its minimum. Generally the turnaround report along with the inspection outlook report should be released within 6 months of the completion of the turnaround.

Preliminary turnaround job list: The preliminary job list is formed based on the turnaround outlook report. The preliminary job list should also comprise inspection reports released after the outlook report.

The preliminary job list should be released at least 24 months ahead of the forthcoming turnaround. The preliminary job list once prepared should be released for inspection, operations, and field maintenance department reviews.

Turnaround review meeting (TR 1) (–24 months): The first turnaround review meeting by the steering committee (chaired by the head refinery) should be held 24 months ahead of turnaround (–24 months). The agenda of the meeting should comprise the following points:

- Finalization of a tentative date and duration of the next turnaround
- Firming up of major replacement and modification jobs
- Identifying the major thrust area of the turnaround such as revamp jobs, reliability issues, maintenance/projects-driven, etc.
- Major turnaround philosophy like unit-wise, staggered, or site-wide
- Approval of turnaround milestone chart
- Review of the turnaround report and learning from the previous turnaround

On-stream inspection (–15 months): The on-stream inspection of all the units, their utilities, and their offsite facilities due for turnaround should be completed. The last such inspection report should be released 15 months ahead of turnaround (–15 months).

Improvement/modification jobs (–24 to –15 months): Major modification/improvement jobs should be finalized along with engineering drawings at least 24 months in advance (–24 months). However, smaller modification jobs may be firmed up along with the engineering drawing 15 months ahead of turnaround (–15 months).

Turnaround review meeting (TR 2) (–12 months): The second turnaround review meeting by the steering committee (chaired by the head refinery) should be held 12 months ahead of turnaround (–12 months). The agenda of the meeting should comprise the following points:

- Review of the progress against action items of the last meeting and overall progress of turnaround planning
- Review of major jobs and readiness for the same
- Review of vendor philosophy and major conditions of contract like liquidated damage (LD) and bonus clauses etc.,
- Review of turnaround philosophy
- Review and approval of the turnaround schedule and costs
- Release of final frozen turnaround job list
- Finalize the exact date and duration of turnaround
- Finalize the action plan for material procurement and contracting

Turnaround review meeting (TR 3) (–6 months): The third turnaround review meeting by the steering committee (chaired by the head refinery) should be held 6 months ahead of turnaround (–6 months).
The agenda of the meeting should comprise the following points:

- Review of the progress against action items of the last meeting and overall progress of turnaround planning
- Resolving any issues pertaining to turnaround
- Review of the timing of the proposed turnaround; situation of turnaround of other companies in and around the country/region; its acceptability from the angles of religious functions, weather conditions, etc.
- Review of major jobs and readiness for the same
- Approval of the allotment of the fabrication spaces to vendors

Administrative issues related to turnaround like

- Gatepass, police, verification, canteen facilities, Employee Provident Fund/Employee's State Insurance Scheme (EPF/ESIC) procedures, etc.
- Freezing the turnaround start date
- Review and approval of the changes requested to the turnaround job list

- Requirement of supervisors and engineers from other departments. Review and approval of the organization of execution teams
- Review of operations plan
- Shutdown and start-up plan
- Permit philosophy

Handover of job list to execution teams: A formal presentation should be made by the respective planners to all the members of the execution teams covering the following aspects:

- Responsibility chart of execution team members
- Job list
- Turnaround schedule including prefabrication work schedule and unit shutdown and start-up plan
- Contracts
- Details of permit philosophy, facilities, administrative requirements, housekeeping, scrap disposal plan, etc.
- Expectations from the execution teams
- Monitoring formats

Prefabrication review meetings: These meetings shall be chaired by GMO and the main objective of these meetings shall be to take stock of the prefabrication work, performance of vendors, and to resolve any issues related with prefabrication.

First meeting (−8 weeks): The meeting shall be 8 weeks ahead of turnaround and the main objective of this meeting will be to ensure:

- Presentation on progress by the respective planner
- Proper mobilization by all vendors
- Review of material availability
- Resolution of issues related to turnaround
- Review of shutdown and start-up plan

Second meeting (−4 weeks): The meeting shall be 4 weeks ahead of turnaround and the main objective of this meeting will be to ensure the following:

- Presentation on progress by the respective planner
- Review of prefabrication work progress, initiation of pre-turnaround site work like scaffolding erection, temporary lines erection, etc.
- Resolution of issues related to turnaround

Third meeting (–2 weeks): The meeting shall be 2 weeks ahead of turnaround and the main objective of this meeting will be to ensure the following:

- Presentation on progress by the respective planner
- Review of prefabrication work progress
- Review of progress of site activities
- Resolution of issues related to turnaround

Performance Criteria

This section describes the criteria for customer satisfaction. Often it points to three criteria: cost, schedule, and quality. The project cannot, for example, cost more than a set amount; specific milestones or red-letter dates must be met; service or product specifications must be addressed. This information allows for meaningful planning and ensures that the project will address key concerns.

What Are the Shutdown/Turnaround Performance Criteria?

1. Shorter shutdown/turnaround duration to increase availability of manufacturing facility
2. Reducing scope by efficient risk-based inspection (RBI)
3. Efficient shutdown/turnaround preparation through:
   a. Thorough scope identification to correct level of detail
   b. Realistic and workable shutdown/turnaround plan
   c. Risk analysis procedure (RAP) executed on critical path activities
4. Efficient shutdown/turnaround budget control mechanism for planned and emergent work scope
5. Leak-free start-ups
6. Effectively managed safety environment through innovative safety plans

3

Planning

"Failure to plan is planning to fail." No single strategy is more important than planning.

Shutdown planning is uniquely challenging and is considerably more challenging than routine running maintenance as the shutdown durations are kept extremely short because of the need to keep the unit down to a minimum period. The compressed time "allowed" by management generally demands high manning intensity, and delayed material deliveries have the potential to cause the shutdown completion date to slip. Hence, it is essential to complete effective planning of the entire shutdown process on schedule and within budget. In fact, there is a very strong correlation between the length of time allocated to shutdown planning and the shutdown's ultimate success.

Planning is the technique of picturing ahead every step in a long series of separate operations and so indicating that for each step routine arrangements suffice to cause it to happen in the right place at the right time. The complexity of the planning function is due to the relative complexity of shutdown execution, which is a time of accelerated activity, with numerous vendors, contractors, and heavy equipment engaged in multiple tasks in close quarters. *There are three major facets to "planning," as it is loosely called: planning, scheduling, and control.* The prime goal of the shutdown planning process is to produce a detailed, overall time-based plan—not merely a work list.

Planning helps

- To eliminate or reduce uncertainty.
- To improve efficiency of the operation.
- To obtain a better understanding of the objectives.
- To provide a basis for monitoring and controlling work.
- To completely define all work required so that it will be readily identifiable to each participant.
- If the task is well understood prior to being performed, much of the work can be preplanned.
- Planning allows the planners to present a clear, well-documented, properly focused understanding of the shutdown work.
- You will come to know what extra skills are required for the work.

Shutdown planning should begin 9–18 months before the anticipated shutdown takes place. Seven areas—procurement, engineering, maintenance, operations, quality assurance, health, safety, and the environment (HSE), security, and administration—will have activities that can be executed during this planning period. To ensure these shutdown planning activities are tracked and completed, they should be included in the master execution schedule.

There are two levels of planning: strategic and tactical.

Strategic management planning: The strategic planning for the next shutdown should start at the conclusion of the current one. Too often, this step is left until 2 or 3 months prior to commencing work, and more often than not concerns only writing job specifications or procuring materials and parts.

Strategic planning comprises those actions undertaken by plant management to establish the vision, goals, objectives, and performance expectations for the shutdown. The purpose of the strategic management plan is to blend the objectives of the shutdown with the goals and business objectives of the organization.

The senior management team must determine when to schedule the next major plant shutdown. Information from marketing, sales, accounting, operations, and maintenance is analyzed and incorporated into the business plan and can provide a long-term strategic plan.

Long-range strategic planning should attempt to forecast maintenance requirements, a minimum of 10 years into the future. It is also important to remember that the plant turnaround procedure is a continuous process.

A practical plant shutdown may be used to eliminate the process fouling condition in one area of the plant. A short shutdown is required to satisfy the regulatory requirement.

Process conditions such as internal fouling and catalyst life may require the plant to be shut down. A risk assessment of acceptable minimum flow rates and product specification will determine when the plant is required to come off-line to correct these process bottlenecks.

Using nondestructive testing (NDT) techniques, inspection of pressure equipments, such as electrical and piping systems, can be carried out and they can be checked for erosion, corrosion, or thermal degradation while the plant is in operation. In many cases, a continuous NDT program is also possible to detect degradation. With a detailed analysis of these data, one can envisage when repair or replacement will be necessary.

Shutdown planning committee: The committee should include the maintenance and production managers, safety representative, procurement or warehouse manager, and be chaired by the plant manager. Alliance contractors should also be invited to participate who may provide information on planning activities.

Treat the shutdown like any other project using classic project management techniques. Applying project management techniques for planning, monitoring, executing, and controlling the progress of the shutdown can delineate various scheduling, resources, and cost questions such as

- Is the amount of work doable within the allotted period?
- What are the critical path jobs for completing the shutdown on schedule?
- Have enough resources (personnel, time, money) been allocated?
- How much do I anticipate the shutdown to cost? Can I stay within budget?

Tactical planning: The tactical planning includes detailed job and scope planning, preparing the estimate, preparing the shutdown organization and detailed planning for quality and safety.

It is now time to undertake detailed planning to ensure that the activities performed in the execution phase of the project are properly sequenced, resourced, executed, and controlled.

The detailed planning starts with a kickoff meeting.

Kickoff Meetings

The typical launch of a shutdown begins with a kickoff meeting involving the major players responsible for planning, including the shutdown manager, section engineers for certain areas of knowledge, subject matter experts (SMEs), and functional leads. There can be multiple kickoff meetings based upon the size, complexity, and time requirements for the shutdown. The major players are usually authorized by their functional areas to make decisions concerning timing, costs, and resource requirements. We will list some of the items discussed in the initial kickoff meeting.

Typically, the meetings involved in turnaround preparation would include, but not be limited to, the following subjects: turnaround policy; plant standards; work list; major task review; project review; inspection review; spares review; safety review; quality review, and site logistics. In addition to these, there will be a large number of ad hoc meetings, convened to resolve specific issues (e.g., the resolution of single design, manufacturing or delivery problems).

Initial discussion of the scope of the shutdown including both the technical objective and the business objective will involve:

- Statement of work that lists the purpose, scope, and deliverables of the shutdown
- The assumptions and constraints
- The shutdown coordinator and his group
- The participants' roles and responsibilities

Job of Planning Team

Typical roles and responsibilities would include, but not be limited to

- Gathering basic data for planning from plant and other necessary personnel
- Producing task specifications and networks for all small tasks
- Specifying bulk work tasks and listing them on control sheets
- Producing material requirement sheets for specified tasks
- Producing equipment and service requirement sheets
- Preparing task books for the execution team
- Liaisoning with other necessary personnel
- Checking completed planning documents and passing them to plant personnel for validation
- Checking critical path networks produced by the planners and putting them into the schedule
- Creating and optimizing the turnaround schedule
- The number of planners is dictated by the amount of work to be done and the time and money available

Planning Step One: Developing the Shutdown Work List

When planning a maintenance shutdown, your goal as the shutdown planner should be to produce a detailed, time-based plan. Development of the shutdown work list should begin immediately after completion of a preceding shutdown. During the period between shutdowns, maintenance technicians should add potential work to the work list as they recognize the work. The operation department should initiate a work request for each job, and the request should be routed to the shutdown planner for evaluation. In addition, the shutdown planner should establish a deadline for submitting work to the work list.

As the shutdown work list is a dynamic document, you should meet regularly with representatives from other departments (i.e., the shutdown planning team) involved in the shutdown to review the work list. The primary purpose of these meetings is to eliminate jobs that may no longer be necessary and possibly add new jobs. It is vitally important that the shutdown work list be kept as short as possible. Keeping the list short is both a means to reduce costs and the primary method for focusing on work that can only be performed during a major outage. All other work is deferred to a time outside of the shutdown window. As the cutoff deadline approaches, the shutdown meetings will result in identification of jobs that are most important and achievable with the available resources and time.

The use of specific, challenging lockdown dates is a concept that may be difficult to accept for some individuals. The 18-month budgetary work list is often misunderstood because it is not always clear how one could know in advance what repairs would be needed.

Shutdown planning begins with specification of the work to be done on each unit operation. This work can include inspection and testing of critical components, such as relief valves, repairs and replacement of damaged or worn-out parts, or a project to improve or expand the capacity of the plant.

Often, the true condition of the equipment is not fully known until it has been cleaned and inspected. There are uncertainties that must be considered during the shutdown planning process and a determination should be made as to the extent to which contingency plans should be developed.

Once a complete scope of work has been developed that includes each of the unit operations, the materials, manpower, and duration of the work can be determined for each of those units. This provides the basis for procuring materials and services, as well as the necessary manpower to complete the work in a reasonable period of time. A key manpower consideration involves work scheduled at other nearby plants. These other plants with their shutdowns will be drawing from the same manpower pool.

Work List

Listing jobs for an upcoming shutdown should begin immediately following the preceding shutdown—simply because many of these future jobs will be based on discoveries made during the equipment internal inspections that have just taken place. Throughout the period between shutdowns, all prospective work should be registered as soon as it is recognized. As each job is identified, it should be written up, a work request generated, and passed to the nominated shutdown planner for evaluation. A cutoff date for submitting work requests should be set. After this deadline, no further jobs should be admitted to the work list without the approval of higher management. The shutdown work list is a dynamic document. It should be regularly reviewed in meetings attended by representatives of all the involved departments. At these meetings, some previously registered jobs may be deleted as no longer necessary, and new jobs added. As the shutdown work cutoff deadline approaches, the minds of job requesters will become more concentrated on what is most important and really needs to be achieved with the finite resources available during the limited period of the shutdown. Electrical, piping, and other tie-ins to existing facilities for new projects and plant additions should never be overlooked when compiling the shutdown work list. It is always extremely worthwhile making provision for these essential future connections to avoid the need for further costly shutdowns.

Work List Meeting

Purpose

The meeting gathers together, and validates, the work requests generated by a large number of people in various departments, creating an approved work list for the turnaround. The detailed work list for discussion should include: preventive maintenance, corrective maintenance, safety and quality initiatives; project work, inspection recommendation, plant cleaning routines, and statutory requirements. If there is a very large or complex work list, the decision may be taken to split it into sections and conduct several meetings rather than a single general one, the work being divided by geographical area, functional unit, or craft discipline. If this is done, there must be at least one final meeting to draw all the sub-lists together into a single list for final validation.

Participants should include

- Turnaround manager
- Planning team
- Plant or production manager
- Maintenance manager (or whoever is responsible for the maintenance function)
- Plant supervisors
- Electric air instrumentation engineers and supervisors

Inspection Review Meeting

The inspection review meeting is held to ensure that all necessary statutory and internally generated inspection requirements, including replacement of static equipments, piping, etc., are identified and defined. *In certain circumstances, such as when past inspection indicates no appreciable deterioration in condition or online monitoring indicates that there is no appreciable deterioration, a case can be made for deferring inspection.*

Major Task Review Meeting

Major tasks need to be properly planned and prepared, because at least one of them will define the duration of the turnaround and any of them, if not properly planned and prepared, may go wrong and become the critical path. The purpose of the major task review meeting is to ensure that large, complex, or hazardous tasks are given the due consideration that their importance merits, so that a relevant specification for the task may be produced.

Project Work Review Meeting

Although project work is performed on the same plant at the same time as the turnaround, it may be controlled by several different departments.

It is therefore vital that it be properly integrated into an overall schedule, identifying interactions with turnaround work and avoiding conflicts. The meeting should ensure that project work, normally planned by the project department, is properly integrated into the turnaround schedule so that all conflicts are resolved and requirements met.

Validated Work Scope

All other aspects of turnaround planning, that is, safety, quality, costs, materials, equipment, and resource requirements, are derived from the work scope. It is not possible to plan against an open-ended work list. Therefore, once the scope has been validated, the work list is frozen on a date agreed by the steering group. Thereafter, any requests for work are authorized by the highest executive authority available and, if it can be justified and is approved, should be placed on a "late work" list. The money, time, and resources required to carry it out are extra to budget and are identified as such. Typically, there will be a steady flow of late work requests throughout both the preparation and execution stages of the turnaround (those arising during the latter stage are termed "emergent work," which is discussed in a later chapter). A validated work scope simplifies planning and preparation; late work complicates them.

Planning the Work Scope

Planning of the shutdown work scope should begin well in advance of the shutdown start date. The shutdown planning team must decide whether jobs on the work list truly belong on the work list. The team should remove from the work list the jobs that maintenance personnel can complete during normal operations.

The primary objective of shutdown planning is to reduce the amount of time the plant is down for shutdown maintenance. To accomplish this objective, you should divide the shutdown plan into the following three stages:

1. Preshutdown
2. Shutdown
3. Postshutdown

The shutdown team should push as many maintenance activities as possible into the preshutdown and postshutdown stages of the plan. It is essential that all departments work together to minimize the shutdown time.

Some shutdown jobs will have tasks that technicians can start before the shutdown begins. You should identify these tasks and schedule them to take place during the preshutdown stage. These tasks should include any preparatory work that could slow progress during the shutdown stage.

The postshutdown stage should include maintenance activities that technicians can complete after production start-up activities begin.

The shutdown planning team should place on the shutdown stage work list any work the team feels technicians cannot accomplish during preshutdown or postshutdown stages. The team should use caution and develop the shutdown "worst case" work list to use for scheduling. This approach will provide reasonable protection against exceeding the planned shutdown time.

Planning the Shutdown Jobs and Tasks

After the team identifies and approves each job for the final work list, you can begin analyzing each job in detail. If you are not familiar with the details of specific jobs, then meet with the maintenance personnel responsible for the jobs or who may have experience with completing the jobs to gather necessary details. You must identify the tasks necessary to complete each job and analyze the estimated labor time, materials, and any special needs necessary for completing the job. You should decide on which jobs to carry out simultaneously or individually. Also, be sure to account for scenarios such as materials that have long delivery periods.

A useful tool for completing your analysis is a maintenance shutdown project management system. Using a project management tool specifically designed for maintenance shutdown and turnaround project planning will result in less time-consuming planning and scheduling activities than using a manual, paper-based planning sheet, or generic project management tool.

Work Breakdown Structure

After the work list is finalized, the shutdown planning provides a systematic approach for breaking it down and assembling the details in an organized, informative format. The *work breakdown structure* (WBS) is the tool for breaking down the shutdown into its component parts. It is the foundation of planning and one of the most important techniques used in shutdown management.

The planning manager accomplishes work breakdown with the help of every available resource. As the work packages are identified, the skills needed for them become apparent, and the team members are identified with these skills to be appointed for execution.

In fact, a WBS is sometimes referred to simply as a *task list*. It turns one large, unique, perhaps mystifying, piece of work into many small manageable tasks.

The work package is the critical level for managing a work breakdown structure. Work packages are natural subdivisions of cost accounts and constitute the basic building blocks used by the contractor in planning, controlling, and measuring contract performance. It is simply a low-level task or job assignment. It describes the work to be accomplished by a specific performing organization or a group of cost centers and serves as a vehicle for monitoring and reporting progress of work. It should be possible for the actual management of the work packages to be supervised and performed by the line managers with status reporting provided to the shutdown manager at higher levels of the WBS.

The planning manager must structure the work into small elements that are

- Manageable, in that specific authority and responsibility can be assigned
- Independent, or with minimum interfacing with and dependence on other ongoing elements
- Integratable, so that the total package can be seen
- Measurable in terms of progress

The WBS is the single most important element in planning because it provides a common framework from which

- The total program can be described as a summation of subdivided elements.
- Planning can be performed for the shutdown.
- The WBS provides a detailed structure to estimate and capture costs for equipment, labor, and materials on each task. It lays the groundwork for developing an effective schedule and good budget plans.
- Objectives can be linked to company resources in a logical manner.
- Schedules and status-reporting procedures can be established.
- Network construction and control planning can be initiated.
- The responsibility assignments for each element can be established.
- Though the statement of work defines scope at the conceptual level, a comprehensive look at a shutdown's scope can be accomplished only with a WBS.
- The tasks on the WBS become the basis for monitoring progress because each is a measurable unit of work. Time, cost, and performance can be tracked.
- The level of detail in a WBS makes it easier to hold people accountable for completing their tasks. A well-defined task can be assigned to a specific individual, who is then responsible for its completion.

Generating the WBS

The best way to generate the WBS is as part of the joint planning session. We describe the steps as we look at different approaches to building the WBS.

Top-Down Approach

The *top-down approach* begins at the goal level and successively partitions work down to lower levels of definition until the participants are satisfied that the work has been sufficiently defined.

Team Approach

When time is at a premium, the planning facilitator will prefer the sub-team approach. The first step is to divide the planning team into as many sub-teams as there are activities at Level 1 of the WBS. Then follow these steps:

1. The planning team agrees on the approach to building the first level of the WBS.
2. The planning team creates the Level 1 activities.
3. A subject matter expert leads the team in further decomposition of the WBS for his or her area of expertise.
4. The team suggests decomposition ideas for the expert until each activity within the Level 1 activities meets the WBS completion criteria.

Geographic: If shutdown work is geographically dispersed, it may make sense from a coordination and communications perspective to partition the work first by geographic location/unit and then by some other approach at each location/unit.

Departmental: On the other hand, departmental boundaries may benefit from partitioning the shutdown first by department and then within the department by whatever approach makes sense. We benefit from this structure in that a major portion of the project work is under the organizational control of a single manager. Resource allocation is simplified this way. On the other hand, we add increased needs for communication and coordination across organizational boundaries in this approach.

Planning Step Two: Identify Task Relationships

The sequence in which detailed tasks—work packages—are performed is determined by the relationship between the tasks.

What is the proper sequence?

Any time a series of tasks is performed, there are sequence constraints; that is, certain tasks must be performed before others. Sequence constraints are governed by the relationships of different tasks.

There are just two basic rules when graphing task relationships with a network diagram:

1. Define task relationships only between work packages. Even though a project might have hundreds of work packages and several levels of summary tasks, keep the sequence constraints at the work package level. Summary tasks, remember, are simply groups of work packages, so it would not make sense to put a task relationship between a summary task and its work package.

2. Task relationships should reflect only sequence constraints between work packages, not resource constraints. Changing a network diagram because of resource constraints is the most common error in building network diagrams.

The fact that there are not enough people or other resources to work on multiple tasks at the same time is irrelevant here.

Calculating the Critical Path

Once you've prepared your final schedule, you'll naturally begin to think about how you're going to maintain control and keep the project on schedule. Now is the time to start thinking about the critical path concept. In other words, as activities are completed ahead of schedule or behind schedule, the critical path will change. If you don't recognize a change in the critical path through your ongoing schedule updates, you could very likely reach a point where you are spending time, money, and resources fighting fires that don't necessarily matter (i.e., not on the critical path) and not addressing your actual problem areas.

Planning Step Three: Manpower Strategic Planning

Choosing the right people for the shutdown team becomes job number one. Needed are specialists who can either lead skilled workgroups or execute jobs on their own. Each unit of work that the specialists will be working on is called a task. A task is sometimes small enough to be carried out by one person. Tasks that are larger require a group of workers. The lead worker from each workgroup serves on the shutdown team. The shutdown team then consists of a mix of specialists who will lead a group of workers.

The process of team selection begins with breaking down the total work into major work packages, or bundles of jobs to be performed to accomplish one piece of work.

When every work package is known, the shutdown manager should make a list of all the work packages.

Determination of the Skills Needed

Having identified the work packages through the work breakdown structure, the shutdown manager has a good idea of the skills required of the shutdown workers. One level of breakdown will not produce enough details to expose the skills needed.

Planning Step Four: Estimate Work Packages

To determine the cost and duration of the shutdown, it is necessary to build a cost and schedule estimate for each work package; this is called *bottom-up* estimating. A lot of information is generated in the estimating process, so it is critical to record it in a systematic manner.

The schedule estimate for a task measures the time from initiation to completion. This estimate is usually referred to as *duration* of a task. When building a schedule estimate, it is important to include *all* the time the task will span.

Planning Step Five: Calculate an Initial Schedule

Calculating a schedule may be one of the most well known, but unappreciated, of all shutdown management techniques. It is the key to establishing realistic schedules and meeting them. The initial schedule is calculated by using the network diagram and the duration of each work package to determine the start and finish dates for each task and for the entire project. The initial schedule represents the combination of task sequence and task duration. It is called an initial schedule because it has not taken into account people and equipment limitations. The next planning step uses the initial schedule as a starting point and balances it against the resources available to the project.

Planning Step Six: Assign and Level Resources

Develop Resource Plan

Immediately after the plan is formed, it is necessary to allocate the resources required to undertake each of the activities and tasks within the shutdown plan. Although general groups of resources may have already been allocated to the plan, a detailed resource assessment is required to identify

- Types of resources (labor, equipment, and materials)
- Total quantities of each resource type
- Roles, responsibilities, and skill sets of all human resources
- Items, purposes, and specifications of all equipment resources
- Items and quantities of material resources

A schedule is assembled for each type of resource so that the shutdown manager can assess the resource allocation at each stage in the shutdown.

Resource Planning and Scheduling for a Maintenance Turnaround

Resource scheduling for a maintenance shutdown, turnaround, or outage in a large plant is a complex activity. Not only must you properly plan, estimate, and code all the equipment and work orders but also the resources to do the job may only be available at certain times and in certain quantities during the outage.

Suppose there is a crane requirement in more than one place during the shutdown. You need to plan accordingly. Organize your activities by equipment: Decide in which sequence the work will take place on the equipment. Estimate the number of resources you will need and the time it will take to accomplish the activity. Check if one equipment's work depends on another's work being accomplished first. Prioritize the equipment and/or activities so that you have a schedule that will accomplish your key timelines and equipment usage and strategies.

1. Decide your resource availability:
   a. Which resources will you need?
   b. How many will you need?
   c. When will you need them? Only for process shutdown or start-up? Weekdays only? Days and/or nights?
   d. Are your resource availability requirements reasonable? Can you get your contractors in and started on time? Can you fit all your required cranes into a small area of the plant? Can your required trades work in a very small area or will they be forced to work in an area that will require them to move large distances to accomplish the activities on their schedule?
2. How long do you plan your turnaround will take?
   a. Will the period include the pre/post-maintenance work?
   b. When do you plan on ramping up and/or reducing maintenance resources?

After you complete the above-mentioned four steps, you can create a resource schedule. The best way to create the schedule is to complete the following steps:

1. Prioritize the work and look at your resource availability.
2. Schedule the work hour by hour:
   a. Start with the highest-priority items that must be finished first.
   b. If the resource for that hour becomes overused, move to the next hour.
   c. Recalculate your priorities each hour, and then try reworking the schedule again.

Working with anything more than a few dozen activities will cause a lot of manual scheduling and therefore headaches. Imagine having to do this with several thousand activities!

The goal of resource leveling is to optimize the use of people and equipment assigned to the project. It begins with the assumption that, when possible, it is most productive to have consistent, continuous use of the fewest resources possible.

In other words, it seeks to avoid repeatedly adding and removing resources, particularly people, time and again, throughout the project. Resource leveling is the last step in creating a realistic schedule. It confronts the reality of limited people and equipment and adjusts the schedule to compensate time. The goal of resource leveling is to optimize the people and equipment assigned to the project—to plan for consistent and continuous use of the fewest resources.

Process of Resource Leveling

It is important to remember how we are defining the term *resources*. Resources are the people, equipment, and raw materials that go into the project. Resource leveling focuses only on the people and equipment; the materials needed for the project are dictated by the specifications.

The leveling follows a four-step process:

1. Forecast the resource requirements throughout the shutdown for the initial schedule.
2. Identify the resource peaks. Use the resource spreadsheet and the resources histogram where there are unrealistic or uneconomical resource amounts.
3. At each peak, delay noncritical tasks within their float. Remember that float is schedule flexibility. Tasks with float can be delayed without changing the project deadline. By delaying these tasks, you are

also filling in the valleys of the resource histogram, that is, moving tasks from periods of too much work to periods when there is too little work.

4. To eliminate the remaining peaks, reevaluate the work package estimates.

For example, instead of having two or three people work together on a task, consider whether just one person could do the work over a longer period of time.

Planning Step Seven: Develop Procurement Plan

One of the planning activities within the planning phase is to identify the elements of the project which will be acquired from external suppliers. The procurement plan provides a detailed description of the products (i.e., goods and services) to be procured from suppliers, the justification for procuring each product externally, as opposed to from within the business, and the schedule for procurement. It also references the process for the selection of a preferred supplier ("Tender Process") and the process for the actual order and delivery of the procured products ("Procurement Process").

Planning Step Eight: Develop Quality Plan

Meeting the quality expectations of the customer is critical to the success of the shutdown. To ensure that the quality expectations are clearly defined and can reasonably be achieved, a quality plan is documented. The quality plan

- Defines what quality means in terms of this project.
- Lists clear and unambiguous quality targets for each deliverable. Each quality target provides a set of criteria and standards that must be achieved to meet the expectations of the customer.
- Outlines a plan of activities that will assure the customer that the quality targets will be met (i.e., a quality assurance plan).
- Identifies the techniques used to control the actual level of quality of each deliverable as it is built (i.e., a quality control plan).

Planning Step Nine: Develop Communications Plan

Prior to the execution phase, it is also necessary to identify how each of the stakeholders will be kept informed of the progress. The communications plan identifies the types of information to be distributed, the methods of distributing information to stakeholders, the frequency of distribution and responsibilities of each person in the shutdown team for distributing information regularly to stakeholders.

Planning Step Ten: Develop Risk Plan

The foreseeable risks are then documented within a risk plan and a set of actions to be taken formulated to both prevent each risk from occurring and reduce the impact of the risk should it eventuate. Developing a clear risk plan is an important activity within the planning phase as it is necessary to mitigate all critical project risks prior to entering the execution phase of the project.

Once the planning is completed, there is a need for a detailed planning and accurate cost estimate. The detailed cost estimate becomes the standard for keeping costs in line.

The full commitment and support of the senior management is vital for the turnaround's success. They must be committed to the following:

- Frozen work lists
- Work for turnaround should be the work that cannot be undertaken at another time
- Releasing staff and resources as needed
- Minimizing additional work
- Creating a work environment where the turnaround planning and execution are shared responsibilities
- Encouraging a team spirit and attitude to understand and help each other

Turnaround Detailed Planning

This is the point at which detailed planning occurs. This stage helps ensure that all necessary work during the turnaround is incorporated and integrated into the plan. This should include all capital work and the operational shutdown and start-up sequences.

Job and Task Planning in Detail

When each job has been agreed and approved as part of the final work list, it must be analyzed in detail.

Each job's constituent tasks have to be identified, the best task execution sequence decided, and each task's content analyzed and estimated for labor time, materials, and special needs. This can be carried out with the aid of a planning computer system, or manually using a job-planning sheet. It is essential that all job influences are identified and careful thought is given to pinpointing the possible concurrent activities which, when executed at the same time, will allow the elapsed time of the shutdown to be compressed.

During this stage, all work list inputs are gathered and assembled while the organization and schedule continue to develop. Continuous maintainability, reliability, and constructability input, combined with reviews of the work list criteria and philosophy, ensure that the scope of work is focused.

All work list proposals must pass through the core team for review and impact assessment. The core team's activities should focus on firming of all turnaround work inputs, which include

1. Risk-based inspection and reliability items
2. Capital works
3. Compliance items
4. HazOp study outputs
5. Operational requirements
6. Process engineering requirements
7. Environmental, safety, and health needs
8. Maintenance requirements
    a. Maintainability, reliability, and constructability reviews
    b. Resolution of conflicting needs
    c. Long lead-time material procurement

The work list should be finalized 12–18 months before the shutdown date; a key role of the core team is to ensure that this happens. The core team/team leader establishes cutoff dates for turnover of capital items that must be adhered to. This ensures an opportunity to demonstrate what influence this work will have on the overall plan. This final work list is the basis for the final budget estimate that should now be ±10%.

Onstream inspection activities close at this time to ensure that the results are interpreted and appropriate action is integrated into the plan.

A work process must be in place to control, review, and, if appropriate, approve additional work that arises after the work list is finalized. This additional work could be an oversight or that which arises during execution.

During this stage, the core team's activities should focus on

- Final work list
- Final plan for equipment cleaning and personnel entry
- Contracting plan in place and all major contracts
- Critical path schedule
- Detailed execution plans
- Detailed safety plan
- Additional work approval process
- Turnover of all capital work
- Completion of materials procurement plan and long lead-time materials ordered
- "What if" scenarios
- Lifting plans
- Mobile equipment requirements
- Final estimate
- Definition of format and frequency of performance and progress reports

Scope Growth of Identified Work List Items

This category includes such items as "replace six trays in a vessel" becoming "replace nine trays." Although this represents additional work, it is not really outside the scope of the work list, and typically some allowance for such contingencies should have been made in the estimates. Scope growth should be documented for cost control purposes but does not necessarily have to pass through the additional work approval process.

Work scope management: It is well known and understood that unless the work scope is complete, managed, and contained, turnarounds and shutdowns will not meet expectations, and turnarounds will run over time and exceed budget. Turnarounds, in particular, and shutdowns are very complex in nature, involving many diverse and varied tasks. There are items that cannot be fully defined prior to the shutdown and allowance must be made to cover these. There will be "surprises" which result in additional work and these must be accommodated. However, above all, the operators and maintenance personnel are reluctant to put all their "wish list" items on the table up front as it escalates the costs. They prefer to put the obvious items forward and retain the rest, putting the latter forward as "safety," "environmental," or similar issues, and insisting this work must be done prior to starting up the units. This culture is destructive and not in anyone's interest. It must be modified if successful turnarounds are to be achieved. Another major concern is when the main job list does not meet the turnaround criteria.

This can happen when major components/projects are overlooked in putting the work scope together. Discussion on what projects and other work should be included in the turnaround must not be an extended process. As a rule, if the work can be done outside the turnaround timing, it should not be included in the job list. A realistic timing for the agreed work scope must be agreed upon. It is detrimental for management to force an unrealistic time outage on those responsible for the turnaround or shutdown. People will do their best if they believe a target is achievable, but will lose heart and perform accordingly when unrealistic timing is forced upon them. The following section provides guidelines for the development of a work scope.

Work scope development guidelines: Top management must have ownership to ensure complete work scopes during front-end planning stages. This will involve a radical change to routine regarding data input procedures. The turnaround organization must be identified and set up early with responsibilities being well defined. Commitment throughout the organization to meet cutoff deadlines must be attained. The shutdown implementation plan must include procedures of work additions and guidelines for work scope items. Management must ensure commitment to this plan by all participants. "Brainstorming" techniques can easily be used to put in place processes to review the work scope with a view to deleting unnecessary work scope items or doing work more expeditiously.

Define tied and untied equipment: The entire refinery flow should be thoroughly examined and at unit level tied and untied units should be defined. In the next level within the unit, the tied and untied equipments should be defined. It is advisable to have the equipment classified into tied (resulting in unit shutdown), semi-tied (resulting in minor throughput loss), and untied (no effect on the plant throughput or safety/environment). This helps in easy identification of unnecessary jobs of untied equipment entering into the turnaround job list. However, there should be a fixed routine for untied equipments and they should be attended to on run based on the schedule. Questions should be asked if in doubt for qualification criteria of a job. Departments only look at the modification from a functional angle and are not allowed to extend the scope of work. Each modification must be approved by the general manager.

Take inputs from all departments: The work scope items need to come from operating, reliability, quality, and inspection departments. Careful consideration needs to be given to ensure the best forum to assemble from all concerned parties so that no area is overlooked. Once a forum is agreed, prepare an introductory document advising about the forum, its purpose, and the expectations, emphasizing their importance. Senior management must endorse this and if possible undertake to present management expectations at the forum. Open discussion should be encouraged and all items should be tabled and noted. It is recommended that at the initial forum discussion, the pros and cons of each item should be discussed.

Once the initial lists are assembled, the items can then be reviewed by the group at subsequent meetings. In reviewing the list, questions that need to be asked are

- Is it a critical item for
  - Safety and environment?
  - Operability?
  - Reliability?
  - Statutory certification?
- Does it need to be done during the turnaround or can it be done anytime?
- Have all the causes of intermittent failures of the units in the past been taken into consideration?
- Have all the project items been included?

Scrutinize the work: The jobs entering the turnaround work list should be scrutinized against the tied and untied classification and work pertaining to untied equipment should be scrutinized with care and only jobs aligning with the turnaround ideology should be added.

Define cutoffs: The cutoff date for each activity should be defined in the initial milestone chart. However, in each of the meetings related with the turnaround, all the inputs should be assigned a cutoff date. Items submitted after this date must be formally approved by the steering committee after looking at the justification, urgency, and effect on schedule.

Work scope containment strategy: The turnaround work scope should be contained at the time of building the work scope, after finalizing the work scope, and during the execution of the work.

Following are the points to look for:

Staggered turnaround: It is best to have a turnaround of fewer units at a time and do a thorough job for the equipments brought down than to have lot of units and do superficial jobs on equipment. If jobs are done in detail and with good quality, they will give longer life to equipment and will in turn reduce the jobs in the next cycle.

Plant operation: The plant, if operated as per design, shall lead to less wear and tear than if operated beyond design. Hence, management should take a call in deciding the way the plant is operated as it will have a direct bearing on the work scope of the turnaround. If there are some design problems, the same should be set right immediately, otherwise this will result in repeated failures and will load the maintenance system.

Modification jobs: Care should be taken that the scope of modifications does not keep expanding. This can happen especially at the stage when the drawings are to be approved by different departments and individuals start exercising their whims and fancies. The drawing must be approved by the initiator who makes sure that the scope of work is in line with his or her requirements; other (technical) returns on investment for each modification should be worked out.

Inspection jobs: The inspection work constitutes the major part of the turnaround work list. Hence, it is imperative to have an experienced inspection team in place that can forecast jobs which are really critical and cannot wait. Playing too safe will lead to unnecessary work and being too casual will lead to surprises in the turnaround. The inspection work list should be backed up with equipment health data and a detailed analysis of the data collected; if required, specialized agencies may be hired.

Maintenance jobs: The maintenance jobs that are normally waiting for material/outages usually tend to get added into the turnaround job list. Also, the untied equipments which are planned as pre-turnaround work in case of delay get pushed to the turnaround and pose a major problem as these have not been considered in the schedule made for the turnaround. There should be MIS on the pending major jobs with maintenance on run and this list should be closely reviewed by the management. In addition, the maintenance work of the untied equipment should be monitored on run as otherwise they will extend into the turnaround.

Work scope—management of change: The turnaround job list should be frozen about a year ahead of the turnaround. The work advised later should be first cleared by the working committee and then formally approved by the steering committee after reviewing the criticality of the job and its consequences on the turnaround plan, schedule, and budget.

Additional work during turnaround: During the turnaround, additional work comes up as the anticipation is never very accurate and maintenance should be geared up for doing some unanticipated work as per the past experience. However, if the work scope containment steps are followed, the unanticipated work during the turnaround will be the least.

Checklist

- Has a cutoff date been decided for the finalization of the turnaround work scope? Has the cutoff date been followed?
- Has a person been made responsible for compilation of the work list?
- Has the planning cycle for the work scope input and finalization been started in time?
- Has the top management taken ownership to ensure the complete work scope during front-end planning?
- Has the turnaround organization been identified and set up early with responsibilities defined clearly?

- Have the brainstorming sessions been conducted to get all the jobs and their economic justification?
- Have all the departments been involved in the work scope preparation?
- Has the detailed inspection report of the previous turnaround been considered for the work scope preparation?
- Has the inspection department issued a work list for this turnaround?
- Have the common facilities such as utilities been covered in the work list in case the whole refinery shutdown is being planned?
- Have the inputs of the operations department been incorporated?
- Have the inputs from the maintenance department (field maintenance, electrical, instrument, civil) been incorporated into the work scope?
- Have the modifications proposed by technology been incorporated into the work scope?
- Have the jobs which could not be done for want of shutdown been included in the work scope?
- Have the tag jobs and steam leak jobs been included in the work scope?
- Has the rotary cell identified the rotary jobs in consultation with field maintenance and been included in the work scope?
- Has it been ensured that operations and maintenance personnel have not retained certain jobs to be put forward later as "safety" or environmental jobs?
- Have the suggestions given by employees which need turnaround for execution been included in the work list?
- Have the unit revamp jobs and other project jobs been included in the work list?
- Have the proper work order numbers been allotted to modification jobs and been included in the work list?
- Has the following information been included in the work list:
  - Material codes for stock items required
  - Purchase order number and location for direct charge items
  - Drawing references
  - Spade list
  - Hydrotesting pressures
  - Name of contractors for various jobs
  - Gasket schedule
  - Size and rating of lines
  - Detail procedure of job

- Have all the jobs in the work list been checked for meeting turn-around criteria?
- Has the work scope been critically reviewed to delete unnecessary items?
- Has a well-managed, contained, and complete work scope for the turnaround been prepared before the cutoff date?
- Has the conflict between budget and actual work list cost been resolved?
- Has the procedure for work scope additions been clearly included in the turnaround maintenance plan?
- Has it been ensured that the modification job scope will not keep expanding?
- Has a proper procedure for approval of modification drawings been established?
- Has the return on investment calculation been done for all the modification jobs?
- Has a realistic execution time been decided for the agreed work scope?

Information Requirements for Planning

Effective maintenance turnaround planning requires reliable information. Gathering and using reliable information to plan your next maintenance turnaround will improve your planning and scheduling efficiency. You can get the information from SAP or any other computerized maintenance management system (CMMS) software. These tools can maintain historical information from previous maintenance turnarounds.

The information you need falls under the following categories:

- Master equipment list
- Work scope and prioritization
- Tools and equipment
- Materials
- Time
- Labor
- Safety
- Quality
- Cost

Master Equipment List

A master equipment list assists you in determining maintenance requirements and creating job plans. For example, knowing where equipment resides in your plant can help you schedule materials and labor for multiple jobs within the same area.

Work Scope and Prioritization

Start a work list well in advance of the turnaround, possibly beginning at the conclusion of the previous turnaround. Review the list periodically and push as much work as possible to the pre- and post-turnaround stages. This approach will reduce the amount of work necessary during the turnaround, which can shorten the turnaround duration and minimize production downtime.

Tools and Equipment

The work to complete determines the tools and equipment necessary for jobs. Proper identification and availability of tools and equipment affects time, cost, and labor to complete jobs.

When planning the supply of tools and equipment, consider possible breakdown and loss of the tools and equipment. You should plan for adequate backup to minimize the effect on the shutdown schedule.

Materials

Each work order needs a bill of materials. To identify necessary materials, review previous turnaround work orders and job plans.

After identifying necessary materials, requisition and purchase the materials with enough lead time to meet your turnaround schedule requirements. To increase efficiency, have materials placed as close to the job site as possible.

Time

You can obtain time estimates from sources such as previous job plans, maintenance engineers, or CMMS. Time estimates will vary depending on work shift length, employee availability, materials, and so on.

Your time estimates should include activity time and lag time. Activity time refers to time necessary for the maintenance crew to perform the actual task; lag time refers to time beyond the crew's control, such as cooldown time for equipment entry.

Labor

Labor estimates include all trades and skills to complete the turnaround work. The labor falls into two categories: full-time personnel and contractors. You should include both in your planning of labor estimates.

As you plan, keep in mind whether the labor is union or nonunion. If union, then adhere to union agreements and assign correct skills or resources to activities. This step will help avoid disputes and possible work interference.

Safety

Safety guidelines are an important part of your planning and scheduling process. These guidelines will require specific equipment and procedures that require additional labor. You need to include this labor in your labor and time estimates.

Quality

The quality of work affects the time and cost of the turnaround. You should provide guidelines that establish the minimum acceptable quality level and testing requirements.

Establishing minimum quality guidelines can improve safety, prevent rework, reduce equipment failure, and increase equipment life. The net result in the long term is reduction in maintenance and safety costs.

Cost

The previous information categories provide the information you need to prepare cost estimates. As you prepare cost estimates, remember there are multiple cost categories such as direct labor, indirect labor, tools and equipment, and supervisory. These categories can also include "hidden costs"; however, proper planning and scheduling with reliable information can reduce these costs.

Planning Is Not a One-Time Event!

It's inevitable that conditions will change during the shutdown period: someone may be pulled off your team, resources may become unavailable when you need them, the business climate may shift, material shortages could occur, and so forth. As conditions change, your original plan must be modified to reflect those changes. The *shutdown plan is a living document* and you should expect to plan throughout the life of your project.

Beware! Common Planning Failures

Failure to plan in sufficient detail: Sometimes shutdown managers make an attempt to plan, but don't do it in enough detail. Failing to plan and schedule project work in enough detail can result in three significant, undesirable effects, which you can avoid by asking the following questions:

1. Will all involved participants readily understand what it includes? You should *describe and define* elements of work in enough detail so that there's no confusion. I've seen many cases where defining work elements poorly has resulted in rework, as people simply misunderstood what was expected.

2. Can you prepare a reasonably accurate estimate of duration and cost? If a reasonably accurate estimate is needed, the work must be broken down to a point where you can estimate its duration and cost with a high degree of confidence. In other words, the less time and effort you put into defining your shutdown, the greater the uncertainty in your estimate.

3. Will you be able to effectively monitor its progress? The principle is quite simple. To make sure that your shutdown is staying on schedule, you'll need to be able to gauge that team members are making progress as expected. The most convenient way to gauge their progress is by observing the completion of relatively small elements of work routinely—typically at your project team meetings. Therefore, elements of work need to be broken down enough that you can verify their completion readily.

Failure to involve task performers in planning: The principle is simple: the people who will be working on your shutdown should be deeply involved in planning their portion of it. There are at least two good reasons for this. First, the planning outputs will undoubtedly be more accurate as the task performers are probably more knowledgeable than you—after all, it's what they do. Second, involving them during the planning stage is likely to make them significantly more willing to participate and more committed to succeeding. People often feel compelled to live up to what they've promised.

Failure to reflect risk and uncertainty in plans: Generally speaking, risk and uncertainty have the *apparent* effect of extending schedules or making things cost more. Yet, many who plan projects do not properly assess, accommodate, or plan for the inherent risk in the shutdown.

Other common problem areas in planning: These develop and unless properly managed can significantly influence the turnaround performance.
Such areas are listed below:

- Late start to planning process
- Lack of input to job lists
- Open job lists—continually moving target
- Lack of understanding of responsibilities
- Poor purchasing systems
- Lack of team alignment
- No familiarization of contractors to the turnaround plan
- Poor management commitment
- Poor worker productivity
- Poor safety/environmental management

4

Estimating

What Is an Estimate?

Estimating is a big part of shutdown planning. To prepare an accurate, thorough shutdown plan, you'll need to estimate many things: how long it will take to do the work, how much the work will cost, the magnitude of the risk and uncertainty involved, what kind of resources will be required, and other aspects of the shutdown.

Without good estimates we have no way of knowing where we are at any point of time, and we have no way of predicting how much the shutdown will cost or how long it will take.

There are two major things that we estimate in a job: one is the cost or the money that will have to be spent to produce it; and the other is the time that the job will take to be completed. Whenever we are doing the estimates, we will not only be estimating the cost of doing the work but also the time that it will take to complete it.

Methods for Estimating Activity Duration

Estimating the duration of shutdown activities involves identifying the activities and sequencing them in the correct order. Then, time estimates are assigned as baseline measurements and used to track the activities to ensure the shutdown is completed on time. Estimating activity duration is challenging. You can be on familiar ground for some activities and on totally unfamiliar ground for others. This process produces an output called an *activity list*, which is an extension of the work breakdown structure (WBS). The activity list should contain all the activities of the shutdown with a description of each activity so that team members understand what the work is and how it is to be executed.

Understanding the Activity Sequencing Process

We've identified the activities and now need to sequence them in a logical order and find out if dependencies exist among the activities. Here's a classic example. Let's say you're going to paint your house, but unfortunately, it's fallen into a little disrepair. The old paint is peeling and chipping and will need to be scraped before a coat of primer can be sprayed on the house. After the primer dries, the painting can commence. In this example, the primer activity depends on the scraping. You shouldn't prime the house before scraping off the peeling paint. The painting activity depends on the primer activity in the same way. You really shouldn't start painting until the primer has dried.

During activity sequencing, you will use a host of inputs and tools and techniques to produce the final outputs, a shutdown network diagram, and activity list updates.

Dependencies

Three of the inputs to activity sequencing are as follows: mandatory dependencies, discretionary dependencies, and external dependencies.

Mandatory Dependencies

Mandatory dependencies, also known as hard logic or hard dependencies, are defined by the type of work being performed. The nature of the work itself dictates the order the activities should be performed in.

Discretionary Dependencies

These are usually process- or procedure-driven, or "best practices" techniques based on past experience.

External Dependencies

External dependencies are those factors that are external to the shutdown. The weather forecast might be an external dependency in the painting job, because you wouldn't want to start painting prior to a rainstorm.

Estimating Activity Durations

Keep in mind that when you're estimating activity duration, you are estimating the length of time the activity will take to be completed, including any elapsed time needed from the beginning to the end of the activity. Activity duration estimating is performed using the following tools and techniques.

Expert Judgment

Activities are most accurately estimated by the staff members who will perform the activities. In this case, expert judgment is used by team members because of their experience with similar activities in the past.

Analogous Estimating

Analogous estimating, also called top-down estimating, is a form of expert judgment. With this technique, you will use the actual duration of a similar activity completed on a previous shutdown to determine the duration of the current activity—provided the information was documented and stored with the shutdown information on the previous shutdown. This technique is most useful when the previous activities you're comparing are truly similar to the activity you're estimating and don't just appear to be similar.

Reserve Time (Contingency)

You might choose to add a percentage of time or a set number of work periods to the activity or the overall schedule. For example, we know it will take 100 h to run a new cable based on the qualitative estimate we came up with earlier. We also know that sometimes we hit problem areas when running the cable.

During the risk-planning process, we'll look at risk management planning, risk identification, qualitative risk analysis, quantitative risk analysis, and risk response planning. At this stage, we need to identify all potential risks that exist for our shutdown, and we need to understand the probability of risk occurrence. We also want to know what the impact to the shutdown or product outcome will be if the risk does occur. Not all risk is bad, and not all risks have negative impacts, but you need to know about them. All risks are caused by something and therefore have consequences. Those consequences will likely impact one or more of the triple constraints.

Risks associated with the shutdown generally concern the shutdown objectives, which in turn impact time, cost, or quality, or any combination of the three.

The purpose for risk management planning is to create a risk management plan, which describes how you will define, monitor, and control risks throughout the shutdown. The risk management plan becomes part of the shutdown plan at the conclusion of the planning process.

Estimating Resource Requirements

The importance of resources varies from job to job. Types of resources include the following:

People: In most cases, the resources you will have to schedule are people resources. This is also the most difficult type of resource to schedule.

Facilities: Shutdown work takes place in different locations. Planning rooms, porta cabin, temporary sheds, and toilets for workmen are but a few examples of facilities that a shutdown requires. The exact specifications as well as the precise time at which they are needed are some of the variables that must be taken into account. The shutdown plan can provide the details required. The facility specification will also drive the schedule based on availability.

Equipment: Equipment is treated exactly in the same way as facilities. What is needed and when drive the activity schedule based on availability.

Money: Accountants would tell us that everything is eventually reduced to dollars, and that is true. The expenses typically include man, machine, material, facilities, and contracts.

Materials: Parts to be used in the fabrication of products and other physical deliverables are often part of the shutdown work, too; for example, the materials needed to replace the column trays.

Resource Planning

There are several questions you need to consider concerning the resources for your shutdown. As we have pointed out, adding resources doesn't necessarily mean that you will shorten the time needed for various activities. Another factor to consider deals with the skill level of the resources.

Estimating Duration as a Function of Resource Availability

Three variables influence the duration estimate of an activity, and all of them influence each other. They are as follows:

- The duration itself
- The total amount of work, as in man hours/days, that will be done on the activity by a resource
- The percent per day of his or her time that the resource can devote to working on it

Estimating Cost

Now that we have estimated activity duration and resource requirements, we have the data we need to establish the cost of the shutdown. This is our first look at the dollars involved in doing the shutdown. We know the resources that will be required and the number of hours or volume of resources needed. Unit cost data can be applied to the amount of resource required to estimate cost.

When doing an estimate, you need to consider a few concepts. The first is that no matter how well you estimate cost, it is always an estimate. One of the reasons that so many shutdowns result in over budget is that people actually believe that they have done perfect estimating and that their baseline estimate is set in stone. Remember that it is always an estimate.

Some techniques that have proved to be quite suitable *for initial planning estimates* are outlined here.

Extrapolating Based on Similarity to Other Activities

Some of the activities in your WBS may be similar to activities completed in other shutdowns. Your or others' recollections of those activities and their duration can be used to estimate the present activity's duration. In some cases, this process may require extrapolating from the other activity to this one, but in any case, it does provide an estimate. In most cases, using the estimates from those activities provides estimates that are good enough.

Studying Historical Data

Every good shutdown management methodology contains a notebook that records the estimated and actual activity duration. This historical record can be used on other shutdowns. The recorded data becomes your knowledge base for estimating activity duration. This technique differs from the previous technique in that it uses a record, rather than depending on memory.

Historical data can be used in quite sophisticated ways. One of the organization has built an extensive database of activity duration history. They have recorded not only estimated and actual durations but also the characteristics of the activity, the skill set of the people working on it, and other variables that they found useful. When an activity duration estimate is needed, they go to their database with a complete definition of the activity and, with some rather sophisticated regression models, estimate the activity duration. They build products for the market, and it is very important for them to be able to estimate as accurately as possible. Again, our advice is that if there is value added for a particular tool or technique, use it.

Seeking Expert Advice

When the shutdown involves a breakthrough technology or a technology that is being used for the first time in the organization, there may not be any local experience or even professionals skilled in the technology within the organization. In these cases, you will have to appeal to outside authorities. Vendors may be a good source using that technology.

Detailed Cost Estimating Tools

As the preparation phase proceeds and more information is obtained on costs, the estimate is refined with an accuracy of about ±10%. In preparing an accurate cost estimate it must be ensured that unit prices are up-to-date and accurate; either contingency allowances, for an estimated amount of emergent work, are included or, if not, the exclusions are clearly stated; all known or estimated nonproductive time is factored in; all assumptions are stated.

Cost estimating has different tools and techniques used to derive detail estimates: analogous estimating, parametric modeling, bottom-up estimating, computerized tools, and other cost-estimating methods. We'll look at each of these inidividually.

Top-down estimates: These are appropriate when we are doing estimates early in the shutdown or at times when relatively inaccurate estimates are acceptable. These estimates are called rough order of magnitude or budget estimates. When the time comes to commit serious money to the shutdown, bottom-up estimates, often referred to as definitive estimates, should be used for jobs whose breakup is difficult to measure or quantify. A top-down estimate is one in which the entire shutdown of a unit or major jobs in the unit are estimated as a whole.

Analogous estimating: Remember that analogous estimating is a top-down estimating technique and is a form of expert judgment. It is a less accurate technique than the others we'll look at.

Parametric modeling: The idea here is to find a parameter, or multiple parameters, that changes proportionately with shutdown costs and then plug that into the model to come up with a total shutdown cost. In order to use this technique, there must be a pattern that exists in the work so that you can use an estimate from that work element to derive the total shutdown estimate.

Parametric estimates are based on some parametric relationship between the cost and the parameter that can be measured for the shutdown. In a parametric estimate, we use some measurable parameter that changes in the same way that cost does, for example, the inch diameter of a pipe. We then find an adjusting factor that will relate the parameter to the cost of the item being estimated.

Parametric and analogous estimating: These methods are used as types of top-down estimates. With the analogous estimating technique, we identify another shutdown or sub-shutdown that will be used as the basis of the estimate, and we scale it up or down to match the shutdown or sub-shutdown we are trying to estimate. If the actual cost of the basis shutdown is collected accurately and if there is a great deal of similarity between the basis shutdown and the shutdown being estimated, these estimates can be quite accurate. For example, a contractor is estimating a shutdown to build 15 houses. In the past, the contractor has built shutdowns with up to 10 houses. He could estimate the cost of the new shutdown by scaling up the 10-house shutdown by 1.5.

Bottom-Up Estimating

The *bottom-up estimating* technique is the opposite of top-down estimating. Here you will estimate every activity or work item individually and then roll up that estimate, or add them all together, to come up with a total shutdown estimate. This is a very accurate means of estimating, provided the estimates at the work package level are accurate. However, it takes a considerable amount of time to perform bottom-up estimating as every work package must be assessed and estimated accurately to be included in the bottom-up calculation.

A bottom-up estimate is one in which the cost estimates are independently made for small individual details of the jobs. These individual cost estimates are then added together or rolled up to make the cost estimates for the entire shutdown. They can also be added up to create estimates for sub-shutdowns within the shutdown. We can think of a bottom-up estimate in terms of the WBS. If we were to estimate the cost of each task at the bottom of the WBS independently and add the individual estimates together, we would have a bottom-up estimate for the shutdown.

Bottom-up estimates for the shutdown are inherently more accurate than top-down estimates. This is intuitively true. It is mathematically true as well. When a large number of small details are individually estimated and added together, there is a chance that the individual detailed estimates will be either high or low. That is, some of the individual details will be underestimated, and some will be overestimated. When we add them together, some of the overestimates will cancel out some of the underestimates. The result is that the total shutdown estimate's accuracy will improve simply by creating more detailed estimates. Of course, as small details are less likely to have forgotten details within them, overall estimates will improve considerably.

Computerized Tools

Shutdown management software can be a useful tool in cost estimating as are spreadsheet programs. Using software can make the job of cost estimating easy and fast.

Although all of these approaches are valid, some will work better than others. The best approach will depend upon factors such as the availability of historical data, estimating the skills of task performers or subject matter experts, and the amount of time available to prepare an estimate. You may want to try more than one approach, and then use your judgment to come up with the best estimate.

What Is the Cost Baseline?

The cost baseline is the part of the shutdown concerned with the amount of money that is predicted to be spent. The three baselines are closely related, and changes to one of them will result in changes to the others. If a change is made in the shutdown scope baseline, either by adding or removing some of the work that is required, the schedule baseline and the cost baseline will probably have to be changed as well. It is foolish for managers to change the shutdown scope without considering the effect on budgets or schedules, yet it is done by shutdown managers all the time.

Some care must be exercised when planning the cost baseline. As this is the basis for our performance measurement system, we must be careful that the budget is shown on the cost baseline at the same point in time where we expect the actual cost to occur. This can be quite a problem. Suppose we show the budgeted expenditure for an item we are purchasing to be on a particular date, say April 15. To the shutdown management planners, this is the date that the shutdown team completes the work that will make the purchase take place. In other words, the shutdown team decides what it needs to purchase and issues a purchase requisition to the purchasing department to actually buy it. The actual expenditure, the time when the money is actually spent, and the time when the company actually issues a check to the company supplying the parts to the shutdown team might be several months later. Because of the time delay, the actual cost will be below the budgeted cost for several months. This will, of course, have the effect of making the performance reports look better than they are. If there are many delays in the reporting of actual costs in the shutdown, the entire shutdown may look much better than it really is. Of course, sooner or later the actual costs are shown. Then the shutdown performance suddenly falls.

As any shutdown will have many risks, some of them will result in extra cost and work, and others will not. If our estimates for probability and impact were correct, at the end of the shutdown the budget for these risks should end up to be equal to the total expected value of all of the risks that actually took place in the shutdown.

This takes care of the identified risks, but what about the risks that come as a surprise and are not anticipated? These risks, the unidentified risks of the shutdown, must be budgeted as well. Unfortunately, the budget for these risks will be a bit of a guess. We can only rely on our past experience with other similar shutdowns to come up with this estimate. These budgets for unknown and unidentified risks are generally estimated very roughly as some percentage of the total shutdown. The percentage should be quite small, however.

Where the budget for unknown risks is higher than a very small percentage of the shutdown cost, there is a chance that the risk identification process was poorly done and that many of the risks that could have been identified were not. In this case, the budget for the unknown risks is frequently inflated as a way of compensating for this lack of identification. A much better solution would be to spend more time identifying the risks that can be identified, evaluating them properly, and setting aside a small amount of budget for the unknown risks.

There are many pitfalls in producing a good estimate for a shutdown. The deliverables may not all be identified, stakeholders may change their minds, shutdown team members may be optimistic or pessimistic, time may be limited, and so forth. If the shutdown is poorly defined, there is not much of a possibility that the cost and schedule estimates are going to be anywhere near what the actual cost and scheduled time for the shutdown will be. Optimistic schedules can cause problems in estimating as well. Stakeholders or management frequently shorten schedules without adding budget to the shutdown. Generally we can look for increases in cost when schedules are shortened. An inaccurate WBS causes work tasks to be missed. When the individual estimates for the tasks are added up to make a bottom-up, definitive estimate for the shutdown, missed work tasks cause underestimation. Understating risks underestimates our cost and schedule estimates as well. Risks that are not identified and identified risks that have the wrong value for their estimated probability or impact cause management reserves and contingency budgets to be misstated. Cost inflation and failure to include appropriate overheads cause erroneous estimates. It is important to recognize wage and price increases that will occur during the shutdown and adjust estimates accordingly.

Estimating Pitfalls

Estimating is difficult. There are many things that can undermine the accuracy or validity of your estimates. Among the most common pitfalls are the following:

- *Poorly defined scope of work*: This can occur when the work is not broken down far enough or individual elements of work are misinterpreted.
- *Omissions*: Simply put, you forget something.
- *Rampant optimism*: This is the rose-colored glasses syndrome described previously, when the all-success scenario is used as the basis for the estimate.
- *Padding*: This is when the estimator includes a factor of safety *without your knowledge*, a cushion that ensures that he or she will meet or beat the estimate.
- *Failure to assess risk and uncertainty*: As mentioned earlier, neglecting or ignoring risk and uncertainty can result in estimates that are unrealistic.
- *Time pressure*: If someone comes up to you and says, "Give me a ballpark figure by the end of the day" and "Don't worry, I won't hold you to it," *look out*! This almost always spells trouble.
- *The task performer and the estimator are at two different skill levels.* As people work at different levels of efficiency, sometimes affecting time and cost for a task significantly, try to take into consideration who's going to do the work.

- *External pressure*: Many shutdown managers are given specific targets of cost, schedule, quality, or performance (and often more than one!). If you're asked to meet unrealistic targets, you may not be able to fight it, but you should communicate what you believe is reasonably achievable.

- *Failure to involve task performers*: It's ironic: an estimate developed without involving the task performer could be quite accurate, but that person may not feel compelled to meet the estimate, as "it's your number, not mine," so the estimate may appear wrong.

Contingency: The Misunderstood Component

There are a number of technical definitions for *contingency*—basically, any time, money, and/or effort added to the shutdown plan to allow for uncertainty, risk, unknowns, and errors.

A powerful combination of your knowledge of the shutdown, your sense of what you don't know, your experiences on previous shutdowns, the documented experiences of countless other shutdown managers, and some good old-fashioned shutdown manager judgment of your own leads you to the conclusion that an estimating shortfall exists.

Cost Budgeting

Once you've done an estimate, you want to go into the cost budgeting phase. This phase is the time when you assign costs to tasks on the WBS. Cost budgeting is actually very formulaic. You take the needed resources and multiply the costs times the number of hours they are to be used. In the case of a one-time cost, such as hardware, you simply state that cost.

Cost budgeting gives the sponsor a final check on the costs of the shutdown. The underlying assumption is that you've got all the numbers right. Usually you'll have the cost of a resource right, but often it's tough to be exact on the total number of hours the resource is to be used. Remember that no matter what, you are still doing an estimate. Cost budgeting is different than estimating in that it is more detailed. However, the final output is still a best effort at expressing the cost of the shutdown.

Cost Control

The second issue deals with how you look at the numbers you're receiving. If you've done a cost baseline, then you'll have some figures against which you can measure your costs. What you're looking for is a variance from the original costs. The two costs you have at this point are your baseline and the actual costs that have occurred during the shutdown. The baseline is the final estimate of the costs on the shutdown. Your job is to look at the two and determine if management action must be taken.

5

Shutdown Contract Management

What Is a Contract?

A contract is an agreement between two or more parties that is enforceable at law. The rules for contracts established by law vary from country to country.

Scope of a Contract

The terms of a contract should define

- *Who is to be responsible for what?*—Who is to be responsible for defining objectives and priorities, financing, innovation, design, quality, operating decisions? Safety studies, approvals, scheduling, procurement, software, construction, equipment installation, inspection, testing, commissioning, and managing each of these?
- *Who bears which risks?*—Who is to bear the risks of investing, site performance, selecting subcontractors, site productivity, delays, mistakes and insurance?
- *What are the terms of payment?*—For design, development, demolition, construction, management, and other services how payment shall be made. These choices are reviewed later in this chapter.

Parties to a Contract

The parties to contracts are in practice variously called:

- Client, customer, purchaser, or employer
- Contractor, vendor, or supplier

In this book "client" and "contractor" are used to mean the two parties to a contract.

Objectives

To be successful the contract needs to be designed to suit its objectives; the client usually designs a contract. The client may have several objectives—for instance:

- To utilize the skills and expertise of contractors' managers, engineers, craftsmen, buyers, and so on for the limited duration of a project
- To have the benefit of contractor's resources such as proven products, licensed processes, materials in stock, and so on
- To get work started quicker than would be possible by recruiting and training direct employees
- To get contractors to take some of the cost risks of the job, usually the risks of planning to use people, plants, materials, and subcontractors economically

The shutdown contract management shall consist of the following points.

Contractors

The contractor options are explored and issues such as the use of *competitive tenders versus the use of a term contractor, or the use of agency labor* are given consideration for different job needs.

Subcontractors

It is very rare that a main contracting company will carry all of the special skills needed to complete and shutdown. Therefore, a subcontractor engagement plan is formulated by the contractor. This specifies what work is to be carried out by subcontractors and what companies will be invited to tender for the work.

Contractor Packages

The total work scope is broken down into packages that can be tendered against by contractors and subcontractors. This brings up one of the critical elements in contracting out work: the level of specification of work required to enable the contractor to perform the work adequately. This leads to the question of contractor competence—that is, their track record on similar work.

Evaluation/Selection

If competitive tendering is to be used, criteria are set for evaluation and selection. The contract manager is involved and sets guidelines for the selection of a shutdown contractor. Some of the criteria are whether the contractor has already executed similar jobs or not, whether they have good financial standing, etc. As far as possible, the issues of how the client and contractor will fit together into a single shutdown organization are also taken into account.

Contractor Mobilization

To promote the aim of having the right people on the shutdown at the time they are needed, a contractor mobilization plan is drawn up. Contractor briefing and familiarization requirements are taken into account in the program, as is the necessity to check the qualifications of key contractor personnel and scarce resources.

Contractor Monitoring

Working on the principle that you must "inspect what you expect," a routine is drawn up for monitoring the contractors' progress toward completion of the event, safety performance, and compliance with quality requirements. This is done on an ongoing basis and for triggering timely remedial action if problems should arise.

Incentive Schemes

Consider incentive schemes for the purpose of encouraging the contractor to focus on the client's objectives. The use of such schemes must be justified (i.e., they must have a clear measurable benefit for the client). The scope of such schemes may be narrow and focused on a single key indicator such as duration or they may be broad-based and include such things as man-hour savings, safety performance, and quality compliance.

Demobilization Plan

A plan is drawn up to detail the timing and other requirements for contractor demobilization when any particular area of work is completed. This involves a system for recognizing when work is actually complete and should embody a mechanism for debriefing key contractor personnel before they disappear from site. The watchword is earliest demobilization commensurate with the effective termination of the event.

Selection of Contractor Work Packages

The main factors influencing the selection of contractors' work packages are

- The work scope and how it is packaged
- The design of the turnaround organization
- The type of contract to be awarded
- The availability of contractors

Because they are interconnected they also influence each other. There are a number of specific options for packaging work, that is, by work type, functional unit, geographical area, or contractor availability. In reality, the most likely situation will be that a combination of the above options is integrated into an overall contractor plan.

Work package is created by different methods as described below.

By Work Type

A given work package is created by grouping similar tasks together and awarding the whole to one contractor to make it more manageable and more economic to perform. The most common packages are as follows (a representative but not exhaustive list):

- Valve overhaul and replacement
- Pump overhaul and replacement
- Pressure vessel inspection
- Re-traying columns
- Catalyst and packing
- Machines overhaul
- Instrumentation
- Electrical work
- Welding (especially coded welding)
- Water washing and cleaning
- Scaffolding (or staging)
- Lagging (or insulation)
- Painting (or protective coating)

The well-defined package forms the work scope for the contractor. Because each package is made up of similar repetitive tasks it is simpler to price, and its execution easier to control. On the other hand, the resulting engagement of a number of contractors can give rise to contractor–contractor conflict.

By Geographical Area

The plant is divided into discrete areas and all of the work (apart from highly specialized or hazardous tasks) in the area is given to one main contractor. While this arrangement can be complicated, due to the mixture of different tasks that the one contractor must perform, the advantage is that, having only one contractor in the particular area, there is no need to manage issues between contractors.

By Functional Unit

One contractor is awarded responsibility for a whole system (e.g., heat exchanger job of a refinery unit) across a number of geographical areas. The obvious advantage is a greater guarantee of resulting system integrity because it is under a unified control. Conversely, if several contractors work on different parts of one functional system, the turnaround manager must provide a coordinator to ensure that all interfaces between contractors are properly managed and no work is left undone.

Contract Types

1. *Lump Sum*: This is typically used with the design-bid-build method of project procurement.

   - A lump sum contract, sometimes called stipulated sum, is the most basic form of agreement between a supplier of services and a customer. The supplier agrees to provide specified services for a specific price. The receiver agrees to pay the price upon completion of the work or according to a negotiated payment schedule. This type of contract is normally exercised for jobs that are difficult to break down for estimation purpose—for example, maintenance of breach lock exchanger, column internal modification, etc.
   - The stipulated sum contract may contain a section that stipulates certain unit price items. Unit price is often used for those items that have indefinite quantities. A fixed price is established for each unit of work.
   - *Requirements*:
     - Good job definition.
     - Contractor is free to use any means and methods to complete the work.
     - Contractor is responsible for proper work performance.

- Work must be very well defined at bid time.
- Fully developed plans and specifications are required.
- Owner's financial risk is low and fixed at the outset.
- Contractor has greater ability for profit.

- *Advantages*:
  - Low financial risk to owner.
  - High financial risk to contractor.
  - Know cost at the outset.
  - Minimum owner supervision related to quality and schedule.
  - Contractor should assign best personnel due to maximum financial motivation to achieve early completion and superior performance.
  - Contractor selection is relatively easy.

- *Disadvantages*:
  - Changes are difficult and costly.
  - Early job start is not possible due to the need to complete the design prior to bidding.
  - Contractor is free to choose lowest cost means, methods, and materials consistent with the specifications. Only minimum specifications will be provided.
  - Hard to build relationship. Each project is unique.
  - Bidding is expensive and lengthy.
  - Contractor may include high contingency within each schedule of value item.

2. *Unit Price*

- In a unit price contract, the work to be performed is broken into various parts, usually by construction trade, and a fixed price is established for each unit of work. For example, painting is typically done on a square foot basis, welding joint rate is given on an inch diameter basis, etc. Unit price contracts are frequently good for agreements with subcontractors also. They are used for maintenance and repair work. In a unit price contract, like a lump sum contract, the contractor is paid the agreed-upon price, regardless of the actual cost to do the work.

- *Requirements*:
  - Adequate breakdown and definition of work units
  - Good quantity surveying and reporting system
  - Sufficient design definition to estimate quantities of units
  - Experience in developing bills of quantities

- – Payment terms should be properly tied to measured work completion
- – Owner-furnished drawings and materials must arrive on time
- – Quantity-sensitive analysis of unit prices to evaluate total bid price for potential quantity variations
- *Time and cost risk are shared*:
  - – Owner is at risk for total quantities
  - – Contractor is at risk for fixed unit price
- *Large quantity changes (>15%–25%) can lead to increase or decrease in unit prices.*
- *Advantages*:
  - – Complete design definition is not required
  - – "Typical" drawings can be used for bidding
  - – Suitable for competitive bidding
  - – Easy for contractor selection
  - – Early project start is possible
  - – Flexibility—scope and quantities are easily adjustable
- *Disadvantages*:
  - – Final cost is not known at the outset as bills of quantities at bid time are only estimates
  - – Additional site staff are needed to measure, control, and report on units completed
  - – Unit price contracts tend to draw unbalanced bidding

3. *One comprehensive contract* (known severally as turnkeyengineer-procure-install-construct, all-in, package deal, design + build contracts) A comprehensive contract makes one contractor responsible for all the work for a project and has the following potential advantages to clients:

- The client's staff can concentrate on shutdown objectives. Their minimum involvement in design choices may result in fewer design changes.
- The contractor has to manage all the relationships between design, procurement, subcontractors, and so on.
- Contractors have the greatest scope for being efficient and therefore economical and profitable.
- Contractors are able to plan all the work as a whole and so offer shorter programs.
- It encourages the development of larger contractors; but a comprehensive contract has the following *potential disadvantages* compared to other types of contract.

- The process of assessing and pre-qualifying contractors, drafting contract documents, obtaining contractors' comments, getting tenders, negotiating, and assessing competing designs before agreeing a contract may take longer and add to costs, but the overall management of the relationship between design, fabrication, and so on remains the same. It is shifted to the contractor and away from the client's control.
- Much of the work is usually subcontracted, so the client has limited or only indirect ability to assess and influence whether it will be completed correctly and on time.
- Few contractors may have the financial strength or ability and willingness to manage all of a large project, and therefore the choice of the contractor may be limited.
- The entire project is in the hands of one contractor, unless it is a joint venture.
- Contractually, the client's control of the contractor's performance depends upon the effect of their contract, so the client formally has only indirect influence through legal rights to order remedial rather than direct managerial power to control the resources being used.
- Many specialist companies prefer to work directly for ultimate users of their products and services. If employed as subcontractors, they tend not to be so well motivated to perform and formally they can be controlled only through the main contractor.

It is possible that the lengthier the span of a contract, the greater is the risk that the client and contractor will concentrate their attention on the initial work and underestimate the time and resources needed for completion. But in some instances, long-term risk gets greater attention.

The extent that advantage is taken of physical possibilities to employ only one contractor for the work for a project therefore depends upon:

The capacity, know-how, quality, and motivation of the contractor

The number of suitable contractors

The relative importance of quality, safety, time, and cost

The ability of the client to manage a comprehensive contract

One comprehensive contract can therefore be appropriate for a "standalone" project not interdependent with other installations or services, and with the risks shared between client and contractor.

If one contractor is to have "performance" responsibilities—that is, to undertake to design and supply a complete project or systems to meet a specification that states the client's requirements—the client logically has to at least agree upon the performance, life-cycle economy, standards, approvals, and all else that matters in design before agreeing to such a contract. If the client may want to make changes to the specification or the contractor's design after concluding the agreement, the contract needs to include provisions for ordering and agreeing these changes, but only to a limited extent.

Parallel Contracts

The potential advantages of dividing work among contractors in parallel are

- A contractor can be chosen as best in expertise or resources for one particular type of work.
- If one contractor fails, the job can be off-loaded to another contractor.

Potential Disadvantages

The disadvantages of parallel contracts could be the overall work, and particularly the interactions between the parallel contracts and the risks of actions by one contractor causing claims from the other contractors.

Parallel contracts are directly managed by clients.

Selection of Contractor

Selection of the right contractor is a very important task for shutdown management. Choosing the proper contractor from numerous applicants who are available today in the market is a complicated problem for clients. In dealing with the long-term assets, it is crucial to select a proper contractor who could ensure the quality of the constructed building. The achievement of this aim largely depends on the efficiency of the performance of the contractor who is selected. Principals and contractors can often have conflicting interests in that the principal is looking to get the job done at the cheapest price but the contractor will generally be trying to maximize profit, which means they may only provide the minimum standard as required to fulfill their contractual obligation. The use of an appropriate selection process can facilitate a better alignment of the principal's contractor objectives.

The typical nonprice factors considered are

- Relevant experience
- Appreciation of the task
- Past performance
- Sustainability

- Technical skills
- Resources
- Management systems

Contractor prequalification makes it possible to admit for tendering only competent contractors. The undertaken decisions demand taking into consideration many criteria, including, among others, experience and financial standing that are often difficult to be quantified. Long-term, performance-based contracting offers many advantages compared to the competitive tendering approach. One of the main benefits is that long-term performance-based contracting reduces both direct and indirect costs. The modeling of multicriteria selection is getting more and more important because of the increasing rate of competitiveness in business. To plan, control, and organize contractor selection in the most efficient way, one must consider the different aspects of business environment. Contractor choice influences the shutdown success.

Selection Factors

Nonprice Factors

The first section of the questionnaire sought to gauge the importance of a number of factors (using a scale of 1 [meaning of little importance] to 5 [very important]).

There are a number of nonprice factors that come up consistently in work regarding the selection of contractors. The following factors were those that frequently occurred and were chosen to be investigated:

1. *Task appreciation/understanding of the scope of the shutdown and possible risks*

   The assessment of this factor involves determining what level of understanding the contractor has regarding what is required of them to complete the shutdown successfully.

   Another aspect of this factor is judging if the contractor is able to identify and mitigate important risk factors to be considered.

2. *Methodology to deliver*

   The methodology the contractor proposes is also an indicator of the contractor's understanding of the shutdown but it is primarily a way for the assessors to judge if the contractors are using an appropriate method to put the processes in place to complete the work in the specified time frame. Obviously, if there is a time overrun, this results in increased costs to both the principal and contractor.

   It follows that a reasonable methodology is a good indicator of the ability of the contractor and likely success of the shutdown.

3. *Current capacity of company to carry out the work*
The capacity to complete the work is a critical factor. If the contractor already has one or more jobs, then the addition of the new job may not work. When considering current capacity, any other tenders that the contractor has submitted but that have not yet been awarded must be taken into account.

4. *Demonstrated relevant past experience of company*
A contractor with experience of jobs of a similar size and type is beneficial to the job as he or she is likely to be more aware of possible risks and has a better general understanding of the job. Another aspect of past experience is how long ago the project was undertaken as this has a negative impact on the relevance of that experience.

5. *Key personnel experience and roles in relation to the execution of the shutdown.*
The experience of the key personnel can have varying influence over the success of a shutdown. The levels of experience held by company personnel at various levels allow for appropriate reviews of decisions and help to minimize risks.

   The owner may recognize in some cases, lack of experience by some personnel who otherwise have good training, skills, and work ethic will be made up for by good company experience and management review.

   There is no doubt that the individual experience of key personnel who will be directly involved with the shutdown can have a significant impact on the successful outcome of a shutdown. The collective knowledge and experience available to the contractor for a specific shutdown as detailed in the contractor's tender submission will obviously be carefully considered by the owner during tender assessment.

6. *Current financial position/viability of company*
The financial capacity of the contractor can be a critical factor in the successful completion of a shutdown. The risk that unexpected shutdown costs may not be able to be funded increases if the contractor is relying on a single contract for his or her total revenue.

   It is common for thresholds to be applied for minimum annual turnover and there must be sufficient working capital to ensure the smooth running of the shutdown. It has been suggested that a contract should not be awarded to a contractor if the value of the work represents more than 50% of their annual turnover or more than 20 times their working capital.

7. *Proposed program of work*
A realistic program of work further demonstrates the contractor's understanding of the requirements of the contract and that the methods being employed are appropriate to the shutdown.

For whatever reason, the principal may have specific needs for the work to be completed by a specific date or during a specific period. The program should indicate how the principal's requirements are to be met in the most efficient manner. Risks to meeting these requirements should be highlighted in the tender submission, including possible cost penalties or escalation caused by restrictive time frames.

8. *Quality assurance*

As a factor for consideration in selecting contractors for work, the quality assurance practices employed by the contractor gain in importance for more complex, higher-value work. To achieve a total quality outcome at the end, each stage needs to be subjected to the necessary quality assurance processes in order to achieve overall success. Therefore, a demonstrated commitment to quality assurance by a contractor should provide a level of confidence to the principal that the complexity will be properly managed and that the risk to the budget is reduced.

9. *Insurance details*

Even with all due attention paid to best practices for occupational health and safety, quality assurance, and any other relevant practice, it is not possible, at least not from a financial perspective, to fully eliminate all risks associated with the shutdown. Acts of god (force majeure), human error, unforeseen equipment or material failure, and even intentional acts of vandalism or theft by unknown parties are potential risks to the success of a shutdown. Failure to adequately insure against such risks can have huge detrimental effects on the shutdown outcome if not properly covered by an appropriate level of insurance.

In the worst-case situation, an uninsured event may result in the failure of a shutdown, thus leaving the principal in a position of having to reestablish a new contract to complete the work.

Appropriate levels of insurance will provide confidence to the principal that if the insured events do occur, the negative impact they may make can be minimized.

Often, appropriate insurances are a mandatory aspect of a tender submission.

10. *National Code of practice compliance*

More and more, government bodies are requiring contractors to comply with the National Code of Practice for the construction industry. Compliance becomes a mandatory factor for particular types of work, including all new work.

Price

Price remains a major factor in the selection of contractors but is just another factor that should be considered in a similar manner to nonprice factors or should it be considered separately from nonprice factors before a final decision is made on the successful tenderer?

In any process, to consider price the aim should be to achieve value for money while remaining within the principal's budget, which can be an issue in itself depending on the robustness of the budget estimate.

Tender Evaluation Practices

Introduction

It is essential to follow tendering processes in place to ensure open and effective competition for work. As is clear from the aim of the shutdown, appropriate use of factors in the selection of contractors for engineering work is an important part of the process, but how they are best used changes depending on various elements.

Careful consideration needs to be applied to the tender assessment methodology to ensure that it is appropriate for the particular work. At some point in this determination, a person well experienced in tendering processes and with a good knowledge of shutdown jobs needs to review the tender assessment methodology to ensure it is appropriate.

In selection for a panel, therefore, the factors that would be of greater importance might include

- Previous relevant company experience
- Experience of key personnel within the company
- Financial standing of the company
- Quality assurance and management systems

Some practitioners might suggest that this prequalification for the particular type of work therefore covers the assessment of the technical or nonprice factors in the selection of contractors. Thereafter, the successful tenderer for a particular shutdown work can be simply selected on price.

This may be appropriate for low-cost, simple work. However, for higher-cost, more complex work, membership on a panel may be considered to be only an indication that a contractor has demonstrated that the "basics" can be met, but some contractors can obviously do the basics better than others. The same factors can be considered in more detail, but with additional nonprice factors such as task appreciation and methodology, when they are placed in the context of a specific job.

Prequalification and Contractor Selection

The prequalification process typically involves a completed prequalification questionnaire (PQQ) and supporting documents and programs. The purpose of the PQQ is to identify those contracting organizations with effective safety management systems with proactive cultures. The completed PQQ should be evaluated by a review panel comprising of a variety of experts from various departments within the company. Areas of expertise represented on the review panel should include the following (Farrow 33):

- *Safety, health, and environmental issues*—looking at culture, safety management systems, regulatory compliance, and safety performance.
- *Quality issues*—evaluating the ability of the contracting organization to ensure the integrity and quality of the service.
- *Financial issues*—ensuring that resources are available to meet the demands, performance standards, and costs.

The effectiveness of the contractor's risk reduction practices should be the basis for contractor safety prequalification criteria. Commonly used contractor SH&E performance criteria include the following:

- *Integration of SH&E on current projects*: The most effective means of evaluating a contractor's SH&E capabilities is to visit a job site to evaluate performance. An interview of the prospective contractor should also be conducted to assess the prospective contractor's corporate safety culture, SH&E knowledge, management skills, and philosophy.
- *References from previous customers*: The owner should talk with previous customers and determine whether or not previous customers were satisfied with the contractor's SH&E performance.

Management of Obviously Low Bids

If a submitted price is considered by the tender evaluation panel to be too low to effectively deliver the stated scope, the tenderer should be advised that it is believed that his or her tender price is low and he or she should be requested to confirm his or her price. To ensure fairness in the tendering process, the tenderer should be given the opportunity to either confirm his or her price or withdraw the tender.

There are, of course, reasons why a tenderer might be in a position to submit a significantly lower bid than other tenderers such as

- New construction methods
- Innovative equipment or processes
- Favorable relationship with suppliers
- Lower overheads through better innovative management procedures

- "Buying a job" to utilize an otherwise idle plant
- "Buying a job" to get into a new market

Subcontracts

In nearly all engineering and construction work, the main contractors employ subcontractors and suppliers of equipment, materials, and services in parallel.

A common principle is that in a main contract, a contractor is responsible to the client for the performance of his or her subcontractors. Practice varies in whether a main contractor is free to decide the terms of subcontracts or has to match their terms "back-to-back" with the main contract, choose the subcontractors, accept their work, and decide when to pap them. It also varies in whether and when a client may bypass a main contractor and take over a subcontract.

The potential advantages of being a subcontractor are

- Being able to concentrate on providing a specialized or local service for many clients and main contractors, each of whom may have only an occasional job or uneven programs of work
- Interfacing with others are the main contractor's risk
- Avoiding investment in common services, which may be provided more economically by the client or main contractor

The potential disadvantages are

- The inability to plan far ahead
- Speed is needed in agreeing to most subcontracts and in establishing relationships with the main contractors
- Payment may be very delayed in the chain from the client, long after incurring the cost of purchases and work, even though pay-when-paid practices are to be avoided
- Formal communications with clients are indirect

The main contractor usually employs many subcontractors in parallel to use their expertise and resources when required but in turn the subcontractors employ others to supply materials and for specific work in a job.

Parallel direct contracts between a client and each specialist contractor are an alternative to one main contract favored by the client, but in turn the specialist contractors are likely to employ subcontractors. In planning the use of a contractor, a client should therefore consider the effects of a hierarchy of contracts on critical activities and on the motivation of subcontractors to perform.

Liquidated Damages Terms in Contracts

A familiar but not necessarily satisfactory instance of transferring a risk is the inclusion in a contract of "liquidated damages" terms which make the contractor liable to pay a specified sum for a specified breach in performance, such as lateness in delivery. This is an alternative to the contractor being at risk for damages at large, and can be positively reinforced by also offering to pay "bonus" amounts for completion on time or for recovery after delay.

The intention of these terms in contracts is to encourage contractors to avoid being late. In practice, their effectiveness may be limited because a contractor who is aware that he or she may be late can calculate whether it is cheaper to lose the sum specified rather than employ extra resources to recover lost time—especially as the latter often requires using resources uneconomically. Such contractual liabilities may also be mitigated or unenforceable because the contractor can show that the work has been affected by the client, third parties, or risks exempted from the contractor's liabilities.

Terms of Payment

Cost is the primary measure used by clients when selecting jobs, assessing risks, comparing prospective contractors' tenders, and reviewing the immediate and longer-term results. Value for money over the life of the shutdown is thus the concern of clients and, in effect, money for value over the life of a contract is the concern of contractors.

Down Payment or Payment for Preliminaries

These terms are intended to induce a contractor to make a quick start to his or her activities because of the incentive of being paid the costs soon after incurring them.

A potential risk to the client is that the value of early payments may be lost if the contractor subsequently fails to complete his or her contractual obligations.

Protection against this can be arranged by requiring the contractor to provide a performance bond before receiving the payment.

Milestone and Planned Progress Payment Systems

Payment to a contractor can be in stages, in a series of payments for achieving defined "milestones" of progress. The word "milestone" usually means that payment is based upon progress in completing what the client wants. Payment for achieving defined percentages of a contractor's program is known as a "planned payment system."

Compared to paying a lump sum after the completion of the entire work, these terms of payment have the potential advantage for a client in that the contractor has an incentive to proceed with work because payment is made sooner after incurring the costs. The incentive can be increased by the achievement of a milestone earning a contractor a "bonus" payment.

A potential advantage to both parties is that the contractor's risks and financing costs should be less. The client has to meet the extra financing cost instead.

The potential disadvantages are

- The milestones or equivalents have to be defined and their achievement proved. To avoid doubts and disputes they need to be defined precisely; the contract and its management are more complex than the payment of one lump sum on completion.
- The contract should state whether payment is due when a stage is achieved ahead of program and what payment is due if one stage is missed but the next is achieved.

Unit Rates Basis of Payment

Unit rate terms of payment provide a basis for paying a contractor in proportion to the amount of work completed. The total contract price is thus based upon fixed rates but changes if the quantities change.

In some contracts, what is called a "schedule of rates" is similar to a bill or quantities in form and purpose, as contractors when tendering are asked to offer rates on the basis of indications of possible total quantities of each item of work in a defined period or within a limit of say +15% change in the quantity of work which the client requires at any time.

The potential advantages to a client of basing payment to a contractor on unit rates are

- Unit rate payment systems provide a flexible basis for changing quantities of work based upon competitive prices from tenderers, provided there are agreed limits to the variations in quantities.
- The tendered rates from competing contractors can be compared to assess whether the contractors have understood the work and how they have allowed for risks.
- A contract can be agreed upon by using approximate quantities and therefore before the design is complete, and payment is made only for actual quantities of work done.

- Payment depends upon progress, in detail, and should therefore motivate a contractor to complete work.
- If large changes are made in the quantity of an item, a new rate can be based upon the tender rate adjusted for the effects of the change in quantity.
- Design changes can be planned using rates to estimate their costs and to choose the cheapest of alternative ways of making proposed changes.

The potential disadvantages are

- Listing all the work in detail for unit pricing (usually in a bill of quantities or schedule of rates) is complex to prepare and measure. Its preparation and administration require time and resources, particularly the services of quantity surveyors or other experts in measuring quantities.
- Complexity in defining and measuring items provides a basis for uncertainty and disputes.
- The tenders for contracts under which payment will be based upon unit rates and remeasurement need to be checked to see if prospective contractors have inserted relatively high rates for the items of work they expect may increase in quantity, and lower rates for those that may decrease.
- Logically, a rate cannot apply to any quantity at any time during a contract, as costs depend on equipments that are employed during continuous work.
- Unit rates provide a facility for the client to make changes that are avoid able, and so allow design and other decisions, which could be final before inviting tenders to be made only provisionally or postponed.
- It does not totally avoid the problem that design and other changes imposed on a contractor can lead to disputes about their effects on costs.
- Schedules of rates for unit of payment for types of work but not quantities or continuity can be certain, but at the risk to clients that tenderers will state rates that allow for very uneconomic use of resources.
- Dividing the work to be done for a project into separate items for pricing and payment purposes should therefore be no more detailed than is cost-effective for achieving the accuracy of payment control required by the client.

Logistics

Establish integrated procurement planning: Acquiring the parts and materials necessary to ensure shutdown success is generally a divided or fractured activity. Maintenance, production, procurement, and engineering have traditionally played a role in "chasing parts." By establishing an integrated material management effort, accountability and systematic updates and reporting can be established in the months leading up to the shutdown. This ensures that all required material is ordered and nothing is misplaced or lost. The procurement effort then becomes integral to logistics management.

Material and equipment management: The objective of materials requirement planning is to ensure that various items of materials, namely, spares/consumables/hardware, etc., required for jobs to be carried out during the turnaround maintenance are made readily available so that no time is lost on this account. The material planning can be classified into two major categories: long-delivery and short-delivery items. The long-delivery items need to be identified at an early stage and action for their funds and procurement needs to be taken so that the material is available on time. The material planning can be categorized into the following steps:

Material requirement planning: The material planning is done on the basis of the finalized work list and the turnaround network. The material requirement for each broken down job is listed and compiled. It is best to calculate the material requirement based on the broken down jobs and the compiled list should be cross-verified with the previous turnaround requirements. The material planned should preferably be reserved against the work order for the said equipment/job in the turnaround hierarchy of work orders. The material planning for turnaround is generally based on:

- Inspection recommendations of the previous turnaround
- Work list for the proposed turnaround
- Preparation of the bill of materials based on isometrics of the line sections under replacement as per on-stream inspection recommendations
  - Spare parts used for the overhaul of each of the equipments based on the consumption in the previous turnaround
  - Consumable items like gaskets, studs, bolts and nuts, oils and greases, welding electrodes, etc. required for the listed jobs as per the consumption in the previous turnaround

- Data on the availability of various items of materials in the local market and their delivery periods
- Data on items of materials required to be imported from foreign sources and their delivery periods

Compilation and classification of the materials: The material requirement is updated in the work list work-wise. This work-wise material requirement is consolidated into two separate lists: stock items and non-stock items. These two lists are grouped into item classes, for example, valves, piping, pipe fittings, tower hardware, trays, etc.

Procurement action: The stock item list is sent to the warehouse for procurement. The long-delivery items are sent well in advance so that the delivery takes place well in time. The material requisites should be done well before the contract PRs preparation. This will distribute the load at purchase and will require lesser follow-up and fewer omissions. The non-stock item PRs also should be staggered and care should taken that they should not clash with stock item and contract PRs.

Managing Vendors and Contractors

Maintenance wants their processes to run efficiently and their repairs and maintenance actions to be effective. Procurement wants things cheap. They do not see reliability or efficiency; they see the lowest bidder regardless of safety record, plant or industry knowledge, supervisory capability, or rework.

Procurement managers are promoted on their ability to get things done at the lowest possible *price*. For procurement specialists, parts vendors, material suppliers, and contractors competing with one another is the best of all possible worlds because it lowers the price of the item or service. Cheaper is not always better.

Key to successful shutdowns is establishing preferred provider relationships early in the planning process. Determine which contractors/vendors have the best record for supplying material on time.

Follow-up: Once the PRs are sent for stock and non-stock items, follow-up should be done for placement of the purchase orders with the vendors for timely delivery. Any coordination with inspection, purchase, finance, design, etc. needs to be done so that the material is delivered in time.

Receipt, inspection, and updating in CMMS: Once the material arrives in the refinery, the updation in computerized maintenance management system (CMMS) is done for stock items by the warehouse. For non-stock items they are delivered and are updated in the CMMS after they are approved by inspection. Follow-up with inspections needs to be done for timely approval of the items. Due care needs to be taken that all items are updated in the system so that they are viewed by the concerned field maintenance staff at any given points of time. It is advisable to update vendor performance in the CMMS as this will give a valuable database for the future and reduce issues related with delivery and quality.

Material coordination meetings: The team of head materials and maintenance should regularly meet for formal material coordination meetings along with the staff from the warehouse, purchase, and maintenance planning. The meeting should start once the stock requirement has been given to the warehouse with a frequency of at least once a week. These meetings should set priorities, and expedite and resolve issues pertaining to placement of purchase orders, delay by vendors, delay in inspection, etc. The material coordination meetings should continue during pre-turnaround and turnaround.

The following checkpoints should be brainstormed before ordering turnaround material for the turnaround:

- How is the follow-up of materials ordered for the turnaround to be carried out?
- Are necessary priorities being given in the follow-up of defects noticed during in-house inspection of materials received for the turnaround and are replacements being organized and received promptly?
- How the material received for the turnaround is to be arranged/ stored in the reserved area so that the spoilage/search element during issue is reduced to a bare minimum? How are rubber items and gaskets to be stored?
- Some items, especially those being imported, may be received at the 11th hour or even during the turnaround period. Have we organized ourselves to get such items cleared from the customs during the turnaround period?
- Do the materials received for the turnaround require to be preserved? If so, which items?
- Has a system been evolved to receive materials for the turnaround after the normal working time of warehouse?
- Have we got the necessary management approval for procuring materials for the turnaround on a crash basis?
- Have we evolved a procedure to draw materials from the warehouse after the normal working hours of the warehouse?
- Have we taken steps for arranging materials from alternate sources during the turnaround?
- Have we evolved a procedure to utilize performance of vendors supplying materials for the turnaround for vendor rating?
- Have we thought of the procedure to deal with surplus materials from the turnaround after the turnaround work is completed?
- Has the inspection recommendation of previous turnarounds been considered for procurement of long-delivery items?

- Has each and every component of the work list been considered for material procurement planning?
- Have the spares been identified for the procurement of all rotary and reciprocating equipment to be overhauled in the turnaround?
- Has the detailed list of consumables such as gaskets, nuts and bolts, greases, etc. for the shutdown been made based on the work list and consumption in the last turnaround?
- Has the requirement of welding electrodes, insulation material, paints, etc. been estimated based on the quantum of jobs?
- Have the availability and delivery period of various items of materials been checked in the local market?
- Has the material to be imported from foreign sources been identified with the delivery period for separate follow-up?
- Has a list of all stock items with the quantity required for shutdown been prepared and sent to materials department for procurement action?
- Have the complete specifications, quality requirements, inspection criteria, certification requirement, and preferred vendor lists been made for each item and sent to procurement?
- Have the turnaround material requisitions been marked "Turnaround" prominently?
- Have the proper work order number and cost center been used on all the material requisitions for proper cost allocation?
- Has the area been identified where all the turnaround material shall be kept and preserved?
- Is the preservation and tagging method for material received for shutdown decided?
- Are the certification requirements from reputed and approved suppliers clearly included in the material specification?
- Is the inspection requirement by the company inspector or third-party inspector clearly written in the material specifications?

Have the following been checked before planning spare procurement for rotary and reciprocating equipment?

- Availability in store
- Consumption in last turnaround
- Any observation in last turnaround
- Any entry in history card
- Advice of vendor specialist
- Has a material coordination team been set up for single-point responsibility?

Logistics Management

Logistics management is the ordering, delivery, movement, and staging of repair parts, replacement material, and construction equipment (basically the hardware of the shutdown) to maximize the efficiency and effectiveness of the shutdown effort. It is directly related to a plant's management strategy through time. The more the advanced planning time given to procuring parts and materials, the greater is the probability of success. You can never escape the necessity for advanced planning, especially in the area of logistics management. Find out early what you need; ensure long lead-time materials and parts arrive early enough for preassembly and staging.

Site logistics is concerned with the reception, location, protection, distribution, and final disposal of all items and services required for the execution of the turnaround. The logistics team consists of storemen, drivers, cleaners, etc. They report to the logistics coordinator and carry out the tasks set by him or her.

Typically, the logistics coordinator's roles and responsibilities include

- Reporting directly to the turnaround manager
- Defining the plant boundaries and the available surrounding land
- Drawing up a site plot plan showing all logistical requirements
- Setting up stores, lay-down, and quarantine areas
- Receiving on site, locating, protecting, distributing, and disposing of
  - Materials, consumables, and proprietary items
  - Tools and equipment
  - Vehicles and cranage
  - Services and utilities
  - Accommodation and facilities
- Controlling hazardous substances
- Providing for the daily needs of all personnel
- Allocating work to the logistics team

Facilities plan: The turnaround is a project involving deployment of a large workforce and regular facilities are not sufficient. There are certain bare minimum facilities that should be provided for the workmen to get optimum output from the work force. The facilities plan should include the following:

- Suitable lunch/dinner/tea points at the site of the turnaround and craftsmen and contractor workmen at the timings announced for the turnaround.
- Arrangement for drinking water and camp toilet facility near each unit undergoing the turnaround.

- Additional suitable area for parking of cars as more vehicles can be anticipated to come in during the period of the turnaround.
- Identification and allotment of areas where contractors may establish their site offices for the turnaround work.
- Identification of fabrication yard facilities for prefabrication.
- Identification and marking of scrap yards.
- Are all the supporting functions fully involved in the turnaround and aware of their roles in the turnaround?
- Are the personnel from general administration identified for the turnaround?
- Is the medical center fully equipped and manned on a round-the-clock basis to cater to large numbers of people working?
- Has the security at the gate been reinforced to enable entry of contract labor without delay?
- Has the vehicular movement in the turnaround area been made clear?
- Are the vehicle parking slots properly marked in the turnaround area?
- Are tea/snacks arrangements in the field area been made for all the company and contract workmen?
- Are canteen services reinforced to cater to large numbers of people working in the turnaround?
- Are the drinking water arrangements, urinals, and camp toilet facilities made near the units planned for the turnaround?

6

Shutdown Organization

What Is Involved in Effective Shutdown Organization?

It involves

- Identifying all key roles and, in particular, defining their responsibilities (i.e., not just your own element)
- Defining crystal-clear terms of reference and accountabilities for all key roles and bodies, for example, steering committee, execution team, etc.
- Defining "ways of working" for your team, describing how you will work with key partners, suppliers, and the customer(s)

Organization can also involve areas such as

- Effective mobilization
- Developing team charters
- Defining a governance structure that covers key responsibilities, accountabilities, authorities, and decision making
- Developing all key controls, specific measures of performance, and metrics that will be used by the entire team

Essentials of a Turnaround Team

The turnaround team comprises the following:

- Turnaround steering committee (TASC)
- Turnaround working committee (TAWC)
- Turnaround coordinator
- Team leader
- Planning team
- Quality team

- Operations team
- Execution team
- Safety team

Building a turnaround team: Turnaround organization is a matrix organization that gets personnel from other departments apart from regular staff on a temporary basis in order to function. These personnel have different functional and administrative reporting. Hence, supervisors have lesser control over them. The senior people from different functions will decide their team from their function based on the quantum of work, turnaround duration, and number of shifts.

Turnaround steering committee: The turnaround steering committee is chaired by the head of the plant, and members comprise the general managers (GMs) of operations, technical, and shutdowns, deputy general managers (DGMs) of all departments, and the head of maintenance planning becomes the secretary of the committee.

Turnaround Steering Committee

The members of this team typically consist of the facility's senior management. This group provides direction and guidance to the core team to ensure that the turnaround meets the business' needs. A more important function of this committee, however, is to ensure that the turnaround's scope and budget are aligned.

Management often appears to set unrealistic budget expectations. Frequently, however, what appears to be a poor decision is, in fact, appropriate based on the information available. The work process must therefore ensure that there is communication between the core team and the steering committee. The committee must keep abreast of the scope of work and current estimate costs of execution.

Better information quality will enhance the discussions and subsequent decisions. This will solve all the budget problems, but without a regular review of scope and budget alignment, there will be little opportunity to resolve any differences.

What is the role of a member of the steering committee—what would you want them to do to help you, what would you want them *not* to do?

You would want them to

- Help the shutdown manager secure resources. So as a shutdown manager you want people on the steering committee who are the ultimate owners of the bodies you will need for the shutdown.
- Bang heads together within their own domains if their people are having difficulty making their minds up.

- Interact with other steering committee members to resolve interdepartmental disagreements.
- Make timely decisions.

But, as suggested earlier, you would not want them to

- Meddle in the detail
- Keep changing their minds
- Play politics (tricky to prevent!)
- Undermine the authority of the shutdown manager

Turnaround coordinator: The turnaround coordinator is appointed for ensuring the turnaround execution is smooth and there are no coordination issues. The turnaround coordinator should be a person who is well accepted by all departments. A good shutdown manager assumes that his or her key people can "run the show." The manager exhibits confidence in those individuals working in areas in which he or she has no expertise, and exhibits patience with people working in areas in which he or she is familiar. A good shutdown manager is never too busy to help his or her people solve personal or professional problems.

- A poor shutdown manager considers himself or herself as indispensable, is overcautious with work performed in unfamiliar areas, and becomes overly interested in work he or she knows. A poor shutdown manager is always tied up in meetings.

Team leader: Team leaders manage the work of their team. Team leading is a good first step on the shutdown management ladder. You will learn much more about managing shutdowns by being a team leader in a large shutdown than you will by managing a small shutdown—you will get experience of many more control processes because many of those processes simply aren't needed in small shutdowns.

Planning team: The lead planner and the planning team must be highly experienced, understand and maximize the use and facility of the planning program that has been selected, and be able to communicate well with all sections/external agencies. And most importantly, the planning head should be an experienced manager, leader, team player, and have effective problem-solving skills, good communication ability at all levels, analytical ability, capacity to work under pressure, and so on.

Quality team: These positions are usually already in place in most organizations and can readily be incorporated into the team. However, due to the need to complete inspections early and expeditiously, it is usual to supplement the inspection team. It is important that such supplements receive sufficient upfront training and this must be put in place as a part of the pre-turnaround activities. The need to coordinate major and minor activities is very important. It is important to only undertake the shutdown work that cannot be completed with the units on-stream.

Operations team: The operating staff should also be formed into teams clearly defining the responsibilities for rapid and safe unit shutdowns and start-ups. They should be trained to undertake pre-close-up inspections. They should also be encouraged to witness pressure tests in collaboration with the inspection/quality groups. Other appropriate tasks can be allocated to operators during the unit downtime.

Execution team: This team is the real practitioner of the turnaround management team. The people who can really get the best out of company and contract resources should be put on this team. The members of this team must

- Understand the contractor
- Be innovative in maximizing the use of resources
- Be able to communicate meaningfully with the other group members, particularly the planners
- Be a leader/manager

Safety team: One of the most important activities during turnarounds is the safety, health and environment (SHE) function. This should be taken care of by the existing SHE personnel. The team should formulate the fire prevention precautions required to be taken during a turnaround; they should chalk out the training program required for the contract workmen, envisage the medical services required in the premises, etc.

Choosing the right people: Choosing the right people for the shutdown team then becomes job number one. Needed are specialists who can either lead skilled workgroups or execute jobs on their own. Each unit of work that the specialists will be working on is called a task. A task sometimes is small enough to be carried out by one person. Tasks that are larger require a group of workers. The lead worker from each workgroup serves on the shutdown team. The shutdown team then consists of a mix of specialists who will either work alone or lead a group of workers.

The process of team selection begins with breaking down the shutdown into major work packages, or bundles of jobs, to be performed to accomplish each piece of work.

Determination of the Skills Needed

Having identified the work packages through the work breakdown structure, the shutdown manager has a good idea of the skills required of the shutdown workers.

The skills necessary for building a house, for example, are clearly implied. A house must have a foundation. Because installing a foundation is a work package that involves placing forms and pouring cement, it requires the skills of masons. Walls also are a necessary element of a house; construction of these walls is obviously a work package for skilled carpenters. A house must be wired for electrical services; wiring is a work package for electricians. Small workgroups of masons, carpenters, and electricians, headed by the lead specialist in each group, will work on the house. Installing window shades is yet another work package, but it is one that does not require a workgroup. A single specialist can do this type of work. This specialist will also serve on the shutdown team, along with the lead mason, electrician, and carpenter.

The shutdown must be broken down into work packages. The shutdown manager accomplishes work breakdown with the help of every available resource. As the work packages are identified, the skills needed for them become apparent, and the shutdown manager can recruit team members with these skills—after first conferring with their supervisors for help in identifying the best prospects for the team. Recruitment is done on a person-to-person basis, with the shutdown manager explaining the process to each recruit.

Shutdown managers are generalists with many skills in their repertoire. They are problem solvers who wear many hats. Shutdown managers might indeed possess technical skills, but technical skills are not a prerequisite to shutdown management.

Your shutdown team will have technical experts, and they are the people whom the shutdown manager will rely on for technical details. There are shutdown managers with many years of experience in the industry who have successfully managed multi-million-dollar information technology shutdowns. This is because shutdown management techniques apply across industries and across shutdowns. Understanding and applying good shutdown management techniques, along with a solid understanding of general management skills, are career builders for all aspiring shutdown managers.

If roles and responsibilities are not clear, many problems will result:

- Poor leadership: ultimate source of authority unclear
- People do not know their responsibilities: people will not do what they need to do to make the shutdown successful
- Not clear who can decide what: slow decision making
- Others' roles are unclear: you don't know who to talk to, poor communications
- Unauthorized shutdowns start: nobody's job to authorize shutdowns
- Lack of accountability: there's little incentive to do things properly

- Resources aren't committed: nobody is held responsible to account for breaking resource commitments
- Unclear objectives: shutdown objectives are not owned by anyone, everyone has their own opinion, moving target, potential failure

So, before any shutdown begins, we must ensure that, among several other things:

- A clear management hierarchy exists for the shutdown
- Each person has a defined and agreed set of responsibilities
- People will be held accountable by their line manager for performing their shutdown role

The list of the shutdown manager's responsibilities is either very long or very short. The short one is that the shutdown manager is responsible for everything. Let's break that into some of its component parts.

Define and get agreement to roles: If halfway through a shutdown it becomes clear that the sponsor doesn't understand his or her role, whose fault is that? Defining roles means defining roles upward as well as downward. Particularly if it is the sponsor's first shutdown, take half an hour to run through with the sponsor what he or she can do to help his or her shutdown succeed. And as mentioned previously, do the same with each member of the steering committee.

Some shutdowns go nowhere because senior managers aren't clear what the shutdown is really trying to achieve. If you suspect this is the case, take them offsite for a day and get them to sort themselves out and hammer out a clear mission statement for the shutdown. In one case that comes to mind, a shutdown had set out to be all things to all men.

Secure resources and fashion them into a team: Getting promises from the steering committee to provide people for the shutdown is one thing. Getting the managers who directly own those people to actually release them is quite another. Having acquired the bodies, you will need to do some team building. This could just be a get-together meeting. But if your team comprises people from information technology (IT), human resources (HR), finance, marketing, logistics, and administration, and there has historically been antagonism between some of these groups. A weekend party may cost a few thousands but this investment in team building could save you a lot of money and a lot of stress later in the shutdown—definitely worth considering for larger shutdowns.

Plan the shutdown: This covers a multitude of activities. The degree to which the shutdown manager personally plans the shutdown will depend upon its scale. If there are three people in the shutdown team, the shutdown manager can easily plan their work. If there are 300 people in the team, the shutdown manager will clearly need to delegate the detailed planning.

Design shutdown control mechanisms: How will you report progress, control change, manage risks, etc.? Every shutdown is different; you will need to design control and reporting mechanisms that suit your shutdown or at least adapt and modify any standard processes that are available within your company.

Manage shutdown scope: If the scope is simply too big for the budget and time scale, the shutdown will fail. If you carve out a doable scope but then let it grow uncontrolled as you go along, the shutdown will fail.

Manage and report progress: This assumes there is progress to report, of course. We will cover tracking, controlling, and reporting in a later chapter.

Be accountable for quality: The shutdown manager's appraisal should depend not only upon meeting dates and budgets but also upon the quality of what is delivered.

Reward and punish: Handing out the occasional small financial award to a team member who has excelled can really boost team morale, if they view it as well deserved. Where does the cash come from? The sponsor, of course. Or put another way, put a line item in the business case headed "team awards and rewards."

Shutdown Manager's Skill Requirement

Let's discuss each of the skills in more detail.

Communication Skills

One of the single most important characteristics of a first-rate shutdown manager is excellent communication skills. Written and oral communications are the backbone of all successful shutdowns. Many forms of communication will exist during the life of your shutdown. As the creator or manager of most of the shutdown communication (shutdown documents, meeting updates, status reports, etc.), it's your job to ensure that the information is explicit, clear, and complete so that your audience will have no trouble understanding what has been communicated.

Organizational Skills

Organizational and planning skills are probably the second most important skills a shutdown manager can possess. Organization takes on many forms. As shutdown manager, you'll have shutdown documentation, requirements information, memos, shutdown reports, personnel records, vendor quotes, contracts, and much more to track and be able to locate in a moment's notice. You will also have to organize meetings, put together teams, and perhaps manage and organize media release schedules depending on your shutdown.

Problem Solving

Take a little time to examine and analyze the problem, the situation caus-ing it, and the solution alternatives available. After this analysis, the shut-down manager will be able to determine the best course of action to take and implement the decision.

Negotiation and Influencing

Effective problem solving requires negotiation and influencing skills. Simply put, negotiating is working with others to come to agreement. Negotiation on shutdowns will be necessary in almost every area, from scope definition to budgets, contracts, resource assignments, and more.

Leading

Leadership and management are not synonymous terms. Leaders impart vision, gain consensus for strategic goals, establish direction, and inspire and motivate others. Managers focus on results and are concerned with get-ting the job done according to the requirements. Even though leaders and managers are not the same, shutdown managers must exhibit the character-istics of both during different times on the shutdown. Understanding when to switch from leadership to management and then back again is a finely tuned and necessary talent.

Team Building and Human Resources

Shutdown managers rely heavily on team building and human resource management skills. Teams are often formed with people from different parts of the organization. These people may or may not have worked together before—so there may be some component of team-building groundwork that will involve the shutdown manager. The shutdown manager will set the tone for the shutdown team and will help them work through the various team-forming stages to become fully functional.

Shutdown managers are often selected or not selected because of their leadership styles.

The most common reason for not selecting an individual is his or her inability to balance the technical and managerial shutdown functions.

- The greater the shutdown manager's technical expertise, the higher is his or her propensity to overinvolve himself or herself in the tech-nical details of the shutdown.

- The greater the shutdown manager's difficulty in delegating technical task responsibilities, the more likely it is that he or she will overinvolve himself or herself in the technical details of the shut-down (depending on his or her ability to do so).

- The greater the shutdown manager's interest in the technical details of the shutdown, the more likely it is that he or she will defend his or her role as one of a technical specialist.
- The lesser the shutdown manager's technical expertise, the more likely it is that he or she will overstress the nontechnical shutdown functions (administrative functions).
- A good shutdown manager performs his or her own problem-solving at the level for which he or she is responsible through delegation of problem-solving responsibilities.
- A poor shutdown manager will do subordinate problem-solving in known areas. For areas that he or she does not know, he or she requires that his or her approval be given prior to idea implementation.
- A good shutdown manager develops, maintains, and uses a single integrated management system in which authority and responsibility are delegated to the subordinates. In addition, he or she knows that occasional slippages and overruns will occur, and simply tries to minimize their effect.
- A poor shutdown manager delegates as little authority and responsibility as possible, and runs the risk of continual slippages and overruns. A poor shutdown manager maintains two management information systems: one informal system for himself or herself and one formal (eyewash) system simply to impress his or her superiors.
- A good shutdown manager finds that subordinates willingly accept responsibility, are decisive in attitude toward the shutdown, and are satisfied.
- A poor shutdown manager finds that his or her subordinates are reluctant to accept responsibility, are indecisive in their actions, and seem frustrated.

Role of Team Leader

Typically, each team leader might look after groups doing similar jobs. What does a team leader do?

Detailed planning: Team leaders produce the task-by-task, day-by-day plan for each of their team members and liaise with other team leaders to ensure the teams' plans interlock. In large shutdowns, a member of the shutdown office may facilitate this plan interlocking but beware, the plans must remain the property of the team leaders: if the plans are the shutdown office's, the team leaders may not be quite so committed to achieving them.

Control and report team's progress: The team leader manages and reports upon the work of his team. We will cover the mechanics of this in the chapter on tracking, controlling, and reporting.

Ensure the quality of the team's output: In much the same way that the shutdown manager will be held accountable for the quality of the shutdown's outputs, the team leader will be held accountable for the quality of his team's outputs.

Technical leadership: Team leaders usually provide technical leadership, guidance, and coaching to junior team members. User team leaders similarly may have greater business experience than their team members. A good team leader will develop the skills of his team members, and support and encourage them.

Later chapters will cover the mechanics of estimating, risk management, planning, tracking, and so on. While in a small shutdown the shutdown manager will do all of this himself or herself, in larger shutdowns the team leaders do much of it.

Role of Team Member

Even the lowliest team member has some part to play in the management of a shutdown:

- Help to estimate the work that will be assigned to them
- Help to plan their own work
- Sign up to their task plan
- Report status of their work
- Be accountable for the quality of their work
- Alert their team leader of problems and issues
- Suggest improvements to the shutdown plan, control processes, etc.

The extent to which a team member can contribute to estimating and planning will obviously depend upon their experience, but the team leader (or shutdown manager in a small shutdown) should involve all team members in planning. This not only exploits their knowledge and experience but also helps get their buy-in and commitment to the plan: it's their plan, not one imposed upon them from on high.

7

Executing the Shutdown

The execution phase of the shutdown is the most important phase. It is characterized by the performance of a large volume of tasks by a large number of people of many skills and disciplines concentrated in a limited space. If the plan was laid out well, all the allocated resources will be pulled together to realize their full potential. However, the shutdown coordinator should not just sit back and watch it all happen. The effective control and coordination of work is also very important.

Execution can be broken down into a number of subphases, as follows:

- Preshutdown work
- Shutting the plant down (removing inventory, decontaminating, cooling, isolating)
- Opening the plant up (physical disconnection of items and removal of covers)
- Inspecting the plant (visual and instrumental examination and report); *nondestructive testing etc.*
- Installation of new items; overhaul of existing items
- Safety review during execution
- Boxing the plant up (final inspection, replacement of covers, and reconnection)
- Daily meeting/communication
- Monitoring and controlling
- Joint integrity management
- Plant testing (pressure tests, system tests, trip and alarm tests)
- Starting the plant up (reconnecting services and reintroducing inventory)
- Plant cleanup and final inspection (removing all traces) of the turnaround
- Risk management

Kickoff Meeting

The kickoff meeting is the first in a series of team meetings. It is critical for team members because it provides a broad introduction to the shutdown. Among the topics covered at the meeting are planning, and the methods used to plan and execute the shutdown. In addition, team members are put through a series of exercises and discussions. At this time, they learn what lies ahead of them.

Everything anyone ever wanted to know about the shutdown is revealed at the kickoff meeting! It lays the groundwork for what lies ahead for the team. Working through the agenda, the shutdown manager explains procedures to be used to execute the shutdown. The team will learn that the shutdown is a collaborative venture. They will all participate in shutdown planning and share in achieving good communication within the shutdown, and they will learn that their expertise will be needed in creating the shutdown plan. They also will learn that in creating the plan, each team member can support the shutdown by going beyond his or her personal knowledge to query subject matter experts within their organization, by asking for help in developing the information they bring to planning meetings. Because it is important that every team member fully understands these points at the beginning of the shutdown, everyone on the team must attend this meeting!

A *preshutdown* walk-through with everyone associated with the shutdown is advised. Check lighting requirements, free and open access to exits, and review any emergency plans. Check for crane activity, which should be especially heeded during a shutdown. If cranes are scheduled, watch out that jobs aren't scheduled under the crane's lift path.

Safety

A turnaround is like a construction job, so extra precautions should be taken. There may be a lot of people in the same area at the same time during a shutdown, so the potential of one group's work interfering with another's is possible. It's a good idea to step through all jobs that will be going on concurrently. Questions of safety during this activity should be asked:

- Have all personnel, including contractors, been briefed on personnel protective equipment requirements?
- Are safety procedures such as lockout, vessel entry, and hot work understood by all?
- Are safety clearances needed for this job?

- Will adequate lighting be available?
- What type of lockout will be used?
- Will blinds be required?
- Will people be working above other people?
- Will two or more groups be working in close proximity?
- Will a crane or other overhead device be used in this area?

Job Status Update

Up-to-date information is the key to shutdown success. Lack of job information during shutdown execution is the root of many shutdown failures. The status of all jobs must be communicated in a timely manner. A formal routine should be developed to report job progress.

1. Each morning, the shutdown coordinator should communicate with the shift coordinators. If a three-shift operation is being employed, the second (or night) shift coordinator should have already briefed the third (or graveyard) shift coordinator. Information provided by the coordinators should include

   a. Delays and problems that occurred during the prior evening.

   b. Which problems were resolved and which were not. Unresolved problems on back shifts end up being handled by the day shift.

   c. Staffing changes that were made during the evening.

   d. The percentage of completion of all the jobs performed during the evening.

   e. A rough estimate of the elapsed time remaining on each job.

2. During the day shift, all problems which may result in a delay in completion time or an increase in shutdown cost should be communicated.

3. At the end of every day shift, the shutdown planner should meet with all the shutdown support personnel and update the shutdown status.

4. Properly close all work orders from the shutdown as they are completed. Don't wait until the end of the shutdown. This will help keep the shutdown cost accounting up to date.

5. Ask "What heavy equipment coordination is required today?"

Daily Schedule

Some shutdown coordinators wrongly wait until the start of each shift to develop a schedule. They feel they won't know enough about the progress of jobs to develop a good schedule. Still others require the maintenance supervisors to develop their own schedule from the updated master schedule.

It is important that individual daily schedules be broken out of the master schedule before each day of the shutdown. It is best if this task is performed by or under the direction of the shutdown coordinator. The resources have already been properly leveled and extra downtime jobs have been added to better use the eight maintenance employees, eight per shift. Two, 12-hour shifts are employed. The evening shift schedule can also be developed ahead of time. Any changes required to the evening schedule prior to the shift change should be minor. The maintenance first-shift supervisor can communicate these changes.

Reporting Status

A shutdown planner should be prepared to report the status of the shutdown at any time. Early on in the shutdown, it should have been determined who the customers or stakeholders are.

The status meetings held with shift coordinators and other shutdown team members will provide the background information for all communications with operations. The shutdown coordinator should verify the statements made during these sessions and take actions to get the shutdown on track. Once these actions are in place, the shutdown coordinator should conduct a session with operations.

Progress with respect to eventual start-up time is usually the most desired information. Any situations that may extend the shutdown should be reported immediately. Any contingency plans and remedies should also be communicated to operations.

The shutdown manager should also report if the shutdown work is progressing quicker than originally estimated. Operations may want to reschedule operators or other line personnel for the earlier date.

An atmosphere of total communication should exist during the shutdown. Other persons involved in the shutdown, such as maintenance workers, may want to know how their efforts are affecting the shutdown. The shutdown master schedule should be displayed for all to see. The current status toward completion should be indicated on the schedule. Any benchmarks or goals that have been reached should be identified clearly.

A Program Evaluation Review Technique (PERT) chart provides the best overall look at shutdown progress. The jobs that are completed are highlighted. Delays and extensions of certain jobs are also shown, along with a progress update.

Tracking Shutdown Costs

Shutdown cost data should also be published and displayed for all to see. There are two common ways to present the data: cost per week and cumulative costs.

Budget versus actual costs is best tracked using the cumulative (S-curve) and weekly charting method.

This chart shows progress made toward reaching the budgeted goal for the shutdown. Both estimated and actual costs are displayed.

In addition, the following points are taken care of during the shutdown:

- Updation in software to identify overall shutdown status
- System of rescheduling of activities with resource allocation based on daily work progress
- Reporting system for daily shutdown monitoring
- Coordination among departments, that is, maintenance disciplines, production, materials, technical services, and administration
- Monitoring and reporting format for maintenance and shutdown jobs
- Preparedness for unplanned/unforeseen jobs
- Record of shutdown execution and completion, the detail execution phases like safety, quality, monitoring and control are described in separate chapters.

8

Preshutdown Job

The preshutdown works comprise just the beginning of the list for a shutdown manager. The list itself is the result of past experiences of many shutdown managers and planners, some of whom experienced problems because of neglecting to take precautions. It is hoped that this list will ensure that similar problems don't happen to you. The following items should be on the shutdown manager's checklist as a preshutdown job. They consist of situations common to almost any shutdown and each should receive at least some attention when planning for, and dealing with, the logistics of a large shutdown.

Pre-turnaround activities: The success of a group of persons at work is enhanced to a great extent when they function as a team. Experience has also proved that the performance of a cohesive group can reach greater heights when the group is guided by a good leader. To achieve the objective, it is important that due importance be given to pre-turnaround activities as any delay in this will directly affect the turnaround schedules.

We will give a road map of the typical pre-turnaround activities:

Formation of execution teams: As per the organization chart developed for the supervisors and the nonmanagement employees, the cross-functional execution teams should be formed covering all units under turnaround. A formal presentation should be made by the respective planners to all the members of the execution teams covering the following aspects:

- Responsibility chart of execution team members
- Overview of job list detailing pre-turnaround work
- Turnaround schedule including prefabrication work schedule, unit shutdown, and start-up plan
- List of contracts detailing the names of vendors with contact numbers
- Details of permit philosophy for the turnaround facilities, administrative requirements, housekeeping, and scrap disposal plan
- Expectations from the execution teams
- Introduction to monitoring formats and fixing responsibilities for progress feedback

Pre-turnaround planning activities: The planning activities during the pre-turnaround period are mostly related with the following:

Rechecking: Planning needs to do a recheck if any action is left out on any job and ensure that the action is taken immediately.

Material follow-up: Most of the materials are delivered by this time; only materials that have got rejected may be on a replacement schedule and need follow-up. The material follow-up should be done meticulously, normally using a computer program.

Long-Delivery Items

A significant part of the planning and preparation of a turnaround is the timely procurement of hundreds, or even thousands, of items—materials, spares, proprietary plant equipment, and so on. Some of these, such as very large components of critical machines or parts, will be on long delivery, for example, a spare part for a compressor or items procured from foreign countries. In order to identify these, it is necessary to analyze the work list as early as possible and make out a list of the items to ensure sufficient time is allowed for the various procurements. Contact all suppliers or manufacturers of items in the Red Zone and, if possible, negotiate a new delivery time, which will remove the item from the zone.

If the item is not delivered until after the start of the turnaround, expedite for its availability before the shutdown starts. If there is a requirement of a catalyst, its availability is to be ensured before shutdown starts.

Contracts: All contract PRs must be processed by this time. If some jobs are hanging in the balance, the same needs to be finalized. PO placements are to be expedited for the left-out items.

Network: The network created needs to be updated with inputs from the vendor interaction for critical activities and major equipments so that the sequence of activities of the network matches with the execution philosophy.

Monitoring: The monitoring activity begins with the start of prefabrication activity: daily progress of the execution teams needs to be monitored; the network needs to be updated; and MIS released to top management as per the procedure approved by the steering committee.

Pre-turnaround execution team activities: The execution teams need to do the following tasks upon formation:

Familiarization: The execution team needs to familiarize themselves with the plant, job list, contract terms, and contractors. The execution team needs to study the network of the unit assigned and discuss the same in detail with the vendor and minutely work out the execution strategy.

Contractor requirements: Complete the gate pass formalities and other formalities required for the vendor to bring in manpower for turnaround work. Ensure that the vendor gets a place to erect the temporary site office. Ensure that the vendor's manpower undergoes the organized safety training. Introduce the safety officer to the vendor safety representative. Make material issue passes for each of the vendors required to get free issue material.

Marking at field: The pipeline section under replacement and the spade location for hydrotesting isolation need to be marked, along with the respective vendors and processes. A good numbering system should be evolved identifying the unit and line number, and on top of it, there should be a bigger sign of the vendor for identification that this is the vendor replacing the line section. Marking of tag numbers on the tag jobs for identifications can be useful for steam leaks, etc. which on isolation of the system cannot be identified.

Material issue: Issue material to the vendor for prefabrication work and any other pre-turnaround requirements. Keep a track of materials issued as the material needs to be reconciled after the job. Issue on loan spades and other items as per contract terms and condition.

Prefabrication: Organize permits for the prefabrication work and start prefabrication work as planned. Give necessary progress updates in the prescribed format to planning so that proper MIS may be prepared and corrective action be taken as required. Check whether the prefabricated lines are punched with the correct numbering system so that the same does not get erased during shot blast cleaning.

Pre-turnaround meetings: Once the prefabrication activity starts and gains momentum, the following pre-turnaround meetings should be planned to monitor progress and solve any issues affecting the prefabrication progress.

First pre-turnaround meeting: This meeting should be organized about 8 weeks ahead of turnaround. The meeting agenda should include the following:

- Presentation on progress by the respective planner; the progress of all vendors has to be monitored very closely, especially the vendor doing the critical activities.
- Review of mobilization by all vendors, especially the vendor doing the critical 1–3 activities. The vendor's mobilization is very critical as this is a preamble to the vendor's performance during turnaround.
- Review of material availability. All material should have come by this time. Only few materials where rejections are faced may be on replacement schedule.

- Resolution of issues related to turnaround. All issues must be resolved in line with the turnaround philosophy of ensuring adherence to quality, duration, etc.
- Review of visual inspection and load test of all lifts, lifting tackles, chain blocks, ropes, chains, and slings to be used during the turnaround.
- Review of fine-tuned shutdown and start-up plan. Review of progress of facilities being provided for the shutdown and start-up as agreed.
- It is good to meet the major vendors doing critical activities and know their constraints and resolve them to expedite.
- Smoking booths should be set up as per agreed plan. Smoking booths are to be provided with press-on cigarette lighter and water receptacle for dumping the cigarette butts.

Second pre-turnaround meeting: This meeting should be organized about 4 weeks ahead of turnaround. The meeting agenda should include the following:

- Presentation on progress by the respective planner; the presentation should highlight areas of concern, poor mobilization by vendors, etc.
- Review of prefabrication work progress. The poor progress should be highlighted.
- Review of progress on temporary facilities for shutdown/start-up.
- Initiation of pre-turnaround site work like scaffolding erection, insulation pocket removal, bolt servicing, etc.
- Resolution of issues related to turnaround. All issues must be resolved in line with the turnaround philosophy of ensuring adherence to quality, duration, etc.
- Decisions to put in place backup arrangement in case some vendor fails to rise to the expectations as later than this would be too late.
- An area-wise turnaround Control Point Office (CPO) is to be established near the units. All log books, shift log book, drawings, and specifications are to be kept here with the unit in-charge. The site control office should have telephones, computers connected to LAN running CMMS, printer, final critical path network on the notice board, etc.

Third pre-turnaround meeting: This meeting should be organized about 2 weeks ahead of turnaround. The meeting agenda should include the following:

- Presentation on progress by the respective planner.
- Review of prefabrication work progress. The prefabrication work will be in the final stages by now.

- Review of progress of site activities as the site activities will constantly grow to peak during turnaround.
- Resolution of issues related to turnaround.
- Decisions required to be taken now have to be carefully thought over and should be fast as the time available is very short.
- Decisions for staggering units where the prefabrication may not be up to the mark should be taken here. Any change in turnaround start date has to be carefully thought about as it puts a lot of pressure of wages on vendors who are performing well and may lead to manpower shedding by them, which will result in poor mobilization during turnaround.

Activities during the week prior to turnaround: During the week prior to the commencement of the turnaround, the following activities are undertaken:

Hot work preparations: Covering of all drain pits with steel plates and isolating the oily water system so that sparks from any welding work may not ignite the oil and spread the same.

Site facilities: Setting up of tea/snack points, drinking water facilities, toilets, etc. These facilities are to be located suitably so that it is not difficult for the workmen using the same.

Decommissioning check list: Operations are to verify that all the items as per the decommissioning checklist are in place to proceed for the shutdown activities. Normally, the following should be checked:

- All flanges are marked as per the spade list.
- Steam hoses and water wash connections are made ready.
- Flushing oil tanks are made full and flushing oil pumps are checked and kept ready for pumping flushing oil into the unit.
- Intermediate tanks for feed to the downstream unit are made full in case the downstream unit is to shut down later.
- Slop tanks are emptied and their material pumped out.
- Oil catcher pumps are checked and kept ready to pump out oil.
- Operations are to check all gas meters and oxygen meters for readiness.
- The shutdown plan is available in the control room and all the shift officers are conversant with the plan.

Manpower deployment: All manpower should be deployed in the respective area and in the respective shifts as per the organization chart on the day the shutdown activities are to begin. Vendor manpower should also be deployed as per plan.

Checklist

- Are the comments of group supervisors incorporated into the schedules?
- Has the review of material planning and procurement been done?
- Has the review of contract planning and award of contracts been done?
- Have the computer schedules been finalized and distributed to the group/shop supervisors?
- Have all the spares and materials received for turnaround been properly reserved and tagged?
- Has the replacement for rejected material been organized?
- Have the items that are expected to be delayed been identified and reviewed?
- Have the substitutes/alternate sources been identified for the items that are anticipated late?
- Have all the major contracts been awarded at least 2–4 months before the turnaround?
- Have the contractors started mobilizing as per the turnaround schedule?
- Have the contractors started making their site offices?
- Have the contractor engineers and group supervisors discussed the methodology to carry out the job?
- Have the contractors prepared their work schedules and discussed with planning?
- Have the pre-turnaround meetings been organized at various stages of turnaround planning?
- Have the visual inspection and certification of in-house lifting tools and tackles been done?
- Has the requirement of cranes, welding machines, mobile compressors, hydrotesting machines, etc. been identified and availability ensured?
- Has the confirmation from the vendor specialist been obtained for their availability as per schedule?
- Have the pre-turnaround jobs such as piping fabrication, civil foundation, scaffolding, etc. been started and completed?
- Has the inspection of items to be done before turnaround been carried out?
- Have the condition and quantity of scaffolding material been checked?

- Have the operations prepared and given the checklist for the decommissioning facilities required?
- Have arrangements for electrical lights, welding supply, covering of drain pits, etc. been made?
- Have all the cable trench excavation jobs been done before the turnaround in order to avoid disruption to the movement of the crane, etc.?
- Have all the tags been provided as per the tag list of operations before the shutdown of units?
- Have the tea and snacks points been prepared as per the location plan?
- Are the temporary urinals and camp toilet facilities provided near the work place?
- Are all units' hot work permits strategy worked out?
- Are all the spades prepared as per the spade list?
- Is the site office established and located near the unit under turnaround?

Turnaround execution: The actual execution of the turnaround starts with following points.

- *Shutdown activities*: The shutting down of the process unit marks the start of the turnaround execution. On the predetermined day, the operating section starts taking steps to shut down the units in line with the written-down procedures. For this shutting down activity, the following activities are undertaken:
- *Spading for positive isolation*: The unit-wise execution teams for the turnaround of the process unit has the spade list and the required spades have already been organized. The concerned supervisor with his or her team of workmen is kept on duty at each process unit. The clearances are taken from operations for insertion of spades as required and the task taken in hand in right earnest.
- *Pumping out/steaming/cooling/flushing*: Operations reduce the feed to the unit slowly in line with the predetermined procedure and carry pumping out/cooling/flushing, and steaming all the equipment and the pipelines with the following objectives:
  - All hydrocarbons and chemicals are to be removed from all the equipment and pipelines in the process unit.
  - Equipment is to be handed over to execution teams in as clean a condition as possible to enable them to carry out the turnaround work smoothly.
  - All work has to proceed as per plan.

Preshutdown

End-of-Run Monitoring of Plant

The end-of-run monitoring of the plant is required for a realistic assessment of the plant condition.

From 2 months before the shutdown, the end-of-run conditions of the unit and critical observations like temperature survey and pressure drop, fouling of exchanger, behavior of column operation, condition of furnace, etc. should be examined. This will facilitate the assessment of the present condition of the equipment and likely damages while making necessary preparatory arrangements, so that there are fewer surprises/extra jobs upon opening of the equipment. However, if process and equipment performances are closely monitored during the end-of-run conditions prior to the shutdown, a more realistic assessment of the quantum of various jobs can be made and necessary actions taken beforehand so that jobs are executed during the ensuing shutdown in a more systematic manner without any extra time and cost overrun. Besides, such a close monitoring system will reveal the extent to which plant operation can be stretched/sustained so that no major problem crops up during the shutdown on opening of the equipment.

Preshutdown: End-of-Run Monitoring of Plant

Sl. No	Monitoring Parameter	Frequency
(A) Equipment: Furnace		
1.	Heat pick up in the convection coils (hydrocarbon) through delta T	Continuously in DCS
2.	Heat pick up in the radiation coils through delta T	Continuously in DCS
3.	Heat pick up in the convection coils (non-hydrocarbon) through delta T	Continuously in DCS
4.	Opening of pass control valves/back pressure control valve	Continuously in DCS
5.	Furnace charge pump's (running) motor current.	4 hourly at field
6.	Delta P across individual furnace pass	Continuously in DCS/fortnightly for local indications.
7.	Temperature of hot air to furnace (for balance draft)	Continuously in DCS
8.	Furnace box temperature, skin temperatures of coils	Continuously in DCS
9.	Flame pattern of burners w.r.t. flame height, smoke inside furnace box, flame impingement, etc.	4 hourly at field
10.	Condition of coil tubes w.r.t. bowing, etc., hangers, guides	4 hourly at field
11.	Performance monitoring of APH, ID/FD fans, Soot blowers, etc.	4 hourly at field
12.	Total fuel consumption	Continuously in DCS

Preshutdown: End-of-Run Monitoring of Plant (continued)

Sl. No	Monitoring Parameter	Frequency
(B) Equipment: Exchanger		
1.	Delta T and delta P across tube and shell side	Monthly
2.	Performance evaluation of coolers, condensers, and re-boilers	Monthly
(C) Equipment: Column		
1.	Yield pattern of various products/streams	Continuously in DCS
2.	Quality of rundowns w.r.t. degree of separation, etc. (gap and overlap)	Shiftwise
3.	Delta P across trays/packing beds	Continuously in DCS
(D) Equipment: Pump		
1.	Discharge pressure	4 hourly at field
2.	Motor current	4 hourly at field
(E) Equipment: Compressor		
1.	Main motor current	4 hourly at field
2.	Suction pressure (for all stages)	4 hourly at field

Tagging and Marking

The preshutdown tagging and marking should be done for reducing time lag.

The unit maintenance engineer along with the production coordinator will visit all the job sites of minor/tag jobs to see and assess the quantum and type of job involved therein. Further, the group will identify and tag all leftover jobs, if any. All tag jobs should be identified with numbered tags. While issuing clearances for such jobs, the same tag numbers should appear on the clearance for easy identification.

Inspection and maintenance area engineers will jointly mark/tag the area/location where insulation needs to be removed for inspection and to carry out recommended repairs/replacement jobs and also to facilitate the hookup of new facilities with existing ones as per project requirement.

Temporary Buildings and Enclosures

Temporary buildings and enclosures are often the direct responsibility and cost of vendors and contractors. It is advisable to review with the contractor(s) the following areas of coverage:

Other temporary facilities:

1. *Temporary cafeteria or eating facility*: Ensure that some provision is made, including vending equipment. Work through the logistics of restocking vending equipment, that is, when it will be done, which supplier will be used, etc.

2. *Temporary first aid*: Large contractors should provide their own licensed EMT or first aid technician along with a facility for primary

care. Obtain written notification from your contractors any time the contractor's employees require such services.

3. *Temporary lighting*: Temporary parking areas used during 24-hour shutdowns should be provided with adequate lighting.

4. T*emporary showers and change rooms*: Some shutdown work may necessitate the need for "clean" and "dirty" change rooms and shower facilities. The need for, and provision of, such services should be determined and administered before the shutdown begins.

5. *Temporary storage*: Storage for material, tools, and equipment should be the responsibility of the vendor or contractor. Security for such storage, and liability if theft or damage occurs, should be determined before any material, tools, or equipment comes on site.

6. *Temporary telephones*: Temporary telephones should be brought into the plant. These should be located in the normal temporary break areas. It is the responsibility of the vendor or contractor to ensure that abuse of this equipment does not occur.

7. *Temporary toilets and water*: Portable toilets and potable water stations should be brought into the temporary structure area. If these facilities are to be staged within the plant or facility proper, it is advisable to arrange ahead of time how and when they will be serviced.

8. *Temporary power*: If an unusually large contracted work force is expected, the utilities to such a camp town may tax existing capacity. Identifying the potential need and providing a temporary source from the local utility is advisable.

Prefabricated Work

This is usually a piping job or structural job or column tray mock assembly and can cause problems or time delay during execution. In this case, the materials or items that make up the fabrication may be issued to the site and prefabrication may be completed before the start of the shutdown. A copy of the prefabrication documentation, that is, isometric or other drawings, welding procedures, pressure testing procedures, etc., may be obtained from the engineering department. If anything on the prefabrication bill of materials is on long delivery, try to expedite it and ensure it reaches the site before the shutdown.

Accommodation and Facilities

Accommodation (offices, rest rooms, stores, etc.) and facilities (canteens, washrooms, toilets, etc.) need to be in position on site and serviceable before the turnaround starts—so the planning, resourcing, and execution of this undoubtedly qualifies as preshutdown work.

Crowd Control

Obtaining an accurate accounting of all outside personnel admitted into a plant on a large shutdown is a challenge. When employees are required to sign in and out at a checkpoint, it creates a bottleneck at change of shift. Sign-in sheets can also be inaccurate as some employees may sign in other's names.

Today, programmable magnetic cards and portable readers greatly speed up the entry and exit process and help ensure accurate crowd control. A magnetic card can be programmed with an employee's name and company as well as other pertinent information. When the card is swiped through the reader, it notes the exact date and time with the card information. The information is then periodically downloaded into a computer where it can be sorted and used to verify contractors' employees. These systems are invaluable as a check against contractors' invoices on time and material jobs.

Contractors' Insurance Certificates

Most companies require minimum liability protection as well as proof of worker's insurance coverage for on-site contractors or other outside services. Most contracting firms obtain this insurance coverage only for the time period needed. The presence of a contractor in your plant earlier in the year does not imply that any coverage obtained for that job is still active. Additionally, if a shutdown extends beyond an earlier expected completion date, insurance coverage that was obtained for the original time period must be extended by the contractor or it will become invalid. A file should be maintained for such certificates to minimize third-party litigation in the event of injuries, deaths, or major damage.

Safety Training

A written contractor safety policy establishes guidelines to be followed for contractors working at the company. The rules established are

- Provide a safe working environment.
- Govern facility relationships with outside contractors.
- Ensure that contractor employees and our employees are trained to protect themselves from all potential and existing hazards.

A contractor safety program should start with prescreening. A comprehensive checklist and questionnaire should be developed to ensure that different rules/potential hazards are known to the contract employees. In this checklist, you'll ask questions about the company's history with regulatory citations/fines, injury rates, workers' compensation rates, written health and safety policies and programs, etc.

For example, in the hazard section, you should plainly state that all hazardous substances to be used by contractors on your site must be approved by the safety and environmental departments before the substances are brought on site. Or, you should state that contract employees are not permitted to lock out your equipment unless one of your employees is present and locks out the equipment too.

Employing Barricades

Barricades should be considered to restrict the movement into, or presence of personnel in, restricted areas. The barricades can consist simply of "barricade tape," or may be as formal as individuals posted as sentries for any of the following situations:

1. To limit entrance to, or egress from, any particular area of the plant or facility for safety reasons.
2. To define travel corridors for contractors to and from their parking lot and break areas into their work area within the site.
3. To protect all personnel from hazardous areas or to minimize access to such areas, and to limit "right to know" training for all temporary personnel.

Emergency Showers and Eyebaths

Extra emergency showers and eyebaths should always be considered when the number of working personnel increases. These units are available on a rental basis with pressurized and temperature-regulated water supplies. The rental company can also be contracted to provide regular, documented inspection and testing. Request copies of such inspections for your own records.

Special Machines

As the technologies of maintenance become sophisticated, they increasingly involve expensive equipment needing more specialization in their application. A large number of specialist companies have emerged to perform the necessary work. These machines can save much time and money, and some of them are

Thermal imaging
Remote cameras, videos, introscopy
Bolt tensioning and torquing
Laser alignment and measurement
Sound wave signature measurement
Metal spraying

Because there are not many companies in any one field, they are normally booked up many months in advance and do not have the flexibility to suddenly change either their deployment or manning levels. It is therefore important that they are contacted as early as possible and given accurate specification of the work required. If possible, they should be asked to come to the site and demonstrate their technology so that one may judge whether they are appropriate to the need. Alternatively, another site where the technology is being applied could be visited to inspect it in action.

Vendors' Representatives

In certain cases, a vendor's representative is required on site for expert advice.

In order to ensure that they will be on site when required and can provide a good service, ask the vendor to visit the site beforehand if required to check the equipment before the turnaround starts.

If the vendor is from overseas, arrange for his visa at an early date. The preparation should be such that there should be no problem during execution.

Preshutdown Action Plan (Typical)

Sl. No.	Action	Department
1.	Flushing schemes of all the equipments to be furnished for arrangement of blinds/hose connection etc.	*Production*
2.	Possible equipments should be released by production before the shutdown. Maintenance activities should be completed before the shutdown	-do-
3.	Assess requirement of utilities for the shutdown and inform the utility department	-do-
4.	All critical fire-prone areas like trenches containing CBD/FLO lines should be thoroughly cleaned before the shutdown to avoid fire hazard	-do-
5.	All the valves for repair/replacement to be tagged before the shutdown and new ones to be arranged	-do-
6.	Blind list should be prepared for each equipment and circuit	-do-

(continued)

Preshutdown Action Plan (Typical) (continued)

Sl. No.	Action	Department
1.	Blinds of different sizes should be arranged and wherever possible be kept in position with tags, for execution of jobs as per requirement of the shutdown	*Maintenance*
2.	All lifting tools/tackles of departmental use should be revisioned and tested, and contractors' tools should be checked prior to the shutdown	-do-
3.	Minimum 20 sleepers (12 in. × 8 in. × 7 in.) for the placement of tube bundle will be required to be arranged by the contractor	-do-
4.	All required test rings should be kept ready after servicing and numbering with equipment tag-wise/interchangeability-wise. Test rings to be shifted near the respective exchangers	-do-
5.	Apart from hydrotesting pumps available with agencies, hand-operated test pumps and motor-driven pump set (for high-pressure application) along with spare belts should be kept ready	-do-
6.	Required scaffolding material including cup-lock type should be collected by agencies before commencement of the shutdown and preference should be accorded to the use of cup-lock scaffolding wherever possible	-do-
7.	Scaffolding arrangement to be provided for thickness survey of process lines wherever feasible as per inspection requirement	-do-
8.	High-pressure air/steam hose connection wherever required should be provided by the first day of the shutdown	-do-
9.	HOT/EOT cranes should be serviced in position	-do-
10.	All drain and vent blinds from positions should be removed in consultation with the production department just before the shutdown	-do-
11.	High-pressure hydro-jetting machine to be shifted to unit block where required exchangers to be hydro-jetted and required power connection to be done in consultation with the electricity department. Moreover, the area should be properly barricaded	-do-
12.	Requirement of aluminum ladders, if any, to be ensured by collecting the same from stores and other areas	-do-
13.	Adequate number of dewatering pumps and water ejectors to be made available in working condition	-do-
14.	Fire hoses, tee branches, and spray nozzles should be arranged and installed before the commencement of the shutdown	-do-
15.	Adequate numbers of steam hoses of standard length should be issued and kept ready before the shutdown	-do-
16.	Brass hammers should be kept ready before the shutdown	-do-
17.	Cleaning and testing of passivation solution equipment to be done (whenever required)	-do-
18.	Adequate number of flanges connected with fire-water hose nozzles to be kept ready	-do-

Preshutdown Action Plan (Typical) (continued)

Sl. No.	Action	Department
19.	Adequate number of flanges with nipples for steam hose connection (diameter 2″, 3″, 4″, 6″, 8″) to be kept ready	-do-
20.	Tags for identifying steam/valve leak to be made ready with number	-do-
21.	Required material to be drawn from store and kept ready at site in consultation with inspection (furnace tubes/bends, tray valves, heat exchanger tubes, etc.)	-do-
22.	Sufficient quantity of taper plugs of brass/MS in different sizes to be kept ready for plugging of exchanger tubes. The sacrificial anode should also be ensured in sufficient quantity for replacement	-do-
23.	Briefing by shutdown engineers regarding their jobs and study and completion of preparatory jobs for the shutdown	-do-
24.	Tarpaulins to be kept ready before the shutdown for water washing of furnace convection tubes with required stitching	-do-
25.	Prefabrication of all piping (process modification/inspection recommended replacement) should be completed by the contractor before the shutdown	-do-
1.	All shutdown-related documents like log books, scope of contracts, list of contractors, flushing schemes, modification/structure drawings, etc. to be kept in the shutdown office	*Planning team*
2.	PERT network and bar charts for critical equipment should be displayed in the shutdown planning office on the first day of the shutdown	-do-
3.	Preparation of the shutdown schedule including material status, equipment test pressures, and list of consumable items to be made available before shutdown	-do-
4.	Required drawings of major equipment should be displayed in the shutdown office	-do-
5.	The shutdown planning engineer should coordinate with respective departments for gate passes/heavy equipment/store opening/erection of the shutdown shade/telephone and light arrangement/PA system, tea and snack supply, and installation of PC respectively	-do-
1.	Temporary power supply points for welding sets should be provided in consultation with the shutdown engineer	*Electrical*
2.	Temporary power supply points for hydro-jetting/machines will be required near heat exchanger cleaning areas	-do-
3.	Arrangement for provision of hand lamps, fixed lights, halogens, and flood lights should be made as indicated by various departments	-do-
4.	Exhaust fans should be kept ready for providing in the manholes, wherever required	-do-

(continued)

Preshutdown Action Plan (Typical) (continued)

Sl. No.	Action	Department
1.	Temporary urinals and lavatory should be provided at convenient locations and cleaning at regular intervals should be ensured. Separate toilets and urinals for ladies to be installed at suitable locations. Deployment of sweepers may also be ensured regularly for cleaning	*Civil maintenance*
2.	All CBD drains and manholes to be covered with sandbags and suitably mud-plugged	-do-
3.	Temporary water lines for drinking water with taps should be provided along with additional cooler for cold water at some convenient location in consultation with the shutdown planning engineer	-do-
4.	Adequate numbers of shot-blasting machines are to be kept ready during the shutdown	-do-
5.	Wooden box coverage for burner blocks of furnaces may be arranged before the shutdown	-do-
1.	First aid box, cold/hot/vessel entry/working at height permits, gas mask, dust respirators, and earplugs should be provided in sufficient quantity by the first day of the shutdown at the shutdown office and availability to be ensured during the shutdown, whenever required	*Fire and safety*
2.	All contractors' personnel should be briefed and trained on owners' SHE policy before allowing them to enter the shutdown area	-do-
3.	Adequate numbers of safety banners/slogans to be displayed for safety awareness	-do-
4.	Parking areas should be earmarked (e.g., IOC Employees, Contractors) in consultation with the production department for smooth movement of fire tender in case of emergency. Parking of vehicles in front of fire–water hydrants to be avoided and the same to be ensured	-do-
1.	Welding procedures/welder qualification test to be completed before the shutdown for a prefabrication/shutdown job in coordination with the shutdown engineer	*Inspection*
2.	Identification of locations for thickness survey area for scaffolding erection/removal of insulation	-do-
3.	Inspection/thickness survey of new materials/fittings to be done. Radiography and SR of prefabricated jobs to be completed to avoid last moment problems of low thickness, etc.	-do-
4.	Thickness survey of cold/hot piping to be taken up in the preshutdown phase to minimize the workload during the shutdown	-do-

9

Inspection of Equipment

Managing equipment integrity is essential to the safe, reliable operation of process plant equipment. The current condition of equipment and remaining life must be evaluated to perform a risk assessment. The results of the inspection are used to assess the current condition, whereas materials degradation models are used to assess the remaining life. Equipment condition depends on the type of material damage, such as corrosion, fatigue, or creep. Inspection is used to quantify this damage. The type of inspection employed must be tailored to the type of material damage that is expected in service. Thus, selection of the inspection technique(s) is based on an analysis of the equipment operating conditions and past experience. In addition, the remaining life evaluation depends on the material damage mechanism expected in future service.

Traditional Inspection Practice: Background

It is widely accepted that traditional inspection methods employed at prescribed fixed intervals for managing integrity of pressure equipment in service are not necessarily conducive to ensure safety and can contribute to high operational costs. Experience shows that a lack of adequate knowledge of applicable damage mechanisms, their causes, and where they occur in each equipment item together with a lack of understanding of their risk profiles have led to unacceptable consequences, because inspections were not targeted to match all active and potential damage mechanisms applicable to an item. It may be worth noting that unnecessary frequent inspections do not necessarily improve safety. The traditional inspection process is generally regarded now as reactive rather than proactive.

Risk-Based Inspection

Risk-based integrity management is rapidly becoming the best and most appropriate technique for determining inspection and maintenance strategies for industry assets. Risk-based integrity management allows you to find an optimal balance between asset care and business risk, hence maximizing your return. Risk-based integrity management, in contrast to other methodologies, focuses on quantifying total business exposure to equipment failure risks. It provides a rational decision-making logic to apply corrosion prevention, inspection, and maintenance resources to those assets that are vital to your business survival.

RBI-Driven Integrity Assurance Process: What Is It?

Risk-based integrity assurance is a proactive process used to understand the equipment risk profile at any particular time during its life cycle and put in place strategies to manage and reduce potential risks. RBI forms a major part of this assurance process as applied to static equipment items of a plant. It is an optimized inspection plan with other mitigating strategies derived from a detailed integrity risk assessment of each equipment item. In a typical chemical plant, the static equipment items referred may include reactors, distillation columns, heat exchangers, various other types of pressure vessels, reformers, boilers, fired heaters, piping, storage tanks, etc.

RBI Assurance Methodology: Outline

This process covers the application of RBI strategies to reliably and economically manage equipment integrity at chemical plant sites. The assessment process relies on the availability of historical operating data and previous inspection history and is therefore best suited to plants where these data can be reliably obtained. It should be noted that RBI is a methodology to optimize inspection activities based on integrity risk assessment. As such, it is a critical process itself, which must be performed comprehensively and competently. The RBI technology used to carry out the RBI team study must be able to reliably assess equipment item risk profile based on identified damage mechanisms. This is essential to the whole process because the evaluation of an optimized inspection interval for an item mainly depends on its risk profile. Satisfying these requirements would help provide the required self-assurance to plant inspection, mechanical, and operations chemical engineers.

API RP 580 presents the principals and minimum general guidelines for implementing and sustaining an RBI program. Much of the technology in its application is described in API RP 581–Risk-Based Inspection Technology.

Implementation of RBI includes the following steps:

- *Identify equipment scope*—selection and listing of the equipment assets to be included
- *Gather equipment data*—identification and collection of the specific data required on the equipment design, operation, process service, inspection history, and repairs. Sufficient information must be collected to permit the following: establishing acceptance criteria for inspections, modeling loss of containment and business consequences, identifying potential damage mechanisms, and modeling the probability of failure (PoF).

- *Identify damage mechanisms*—identification of known and potential damage mechanisms, and estimation of damage rates. With the focused approach in RBI, it is important that all potential damage mechanisms be considered and included for inspection. A clear understanding of the damage mechanisms and rates is important in developing accurate modeling of the PoF.

- *Determine probability of failure*—the PoF is based on a model for specific damage mechanisms. For example, with a damage mechanism like wall thinning, the model uses the time in service (age), the corrosion rate, initial wall thickness, and information from prior inspections (number of inspections and inspection confidence). API RP 581 is a basic source for these models. Models are also available for cracking and other damage mechanisms.

- *Determine consequence of failure*—this is typically done utilizing simplified dispersion modeling of a variety of hazardous-fluid-release types (toxic, explosion, fire, etc.) for safety consequences. The intent of these "dispersion" models is to provide a rough (order of magnitude) result as a basis for relative risk ranking of the equipment. Business impact consequence (downtime, loss of production, cleanup, etc.) can also be included.

- *Inspection planning*—planning selects the inspection techniques suited for the potential damage mechanisms and the risk levels to determine the extent (i.e., full or partial) and frequency of the activities. Guidance on the techniques comes from various API or other practices. Guidance on the extent and frequency comes from *inspection strategies* or guidelines developed by subject matter experts or from running various cases and evaluating the impact on future risk versus user-defined risk tolerances.

  - *Perform inspections*: Use qualified inspectors to properly perform and document the inspections to the planned extent and frequency.

  - *Evaluate and update*: Evaluate inspection results and update information on equipment condition (e.g., current wall thickness). Damage rates are reviewed and updated. Nonconformances are evaluated and classified for appropriate follow-up (e.g., reinspection); deficiencies are identified and set up for correction.

  - *Address deficiencies*: If any deficiencies are identified, appropriate follow-up for correction is made (e.g., fitness for service analysis, repair, replacement, etc.).

The new inspection data and updated age are used to recalculate the probability of failure and risk ranking for use in the next round of inspection planning. Addressing issues and other opportunities for corrections and improvements should be part of the ongoing work process. Based upon experience, the inspection frequency of all the equipment is included to minimize the job load in a shutdown. This frequency can be modified based on further observations.

Role of Nondestructive Testing (NDT) in RBI-Driven Integrity Assurance Process

When the RBI technology process is used to justify extended inspection intervals, in particular for high-consequence items, or extended plant run-length times between turnarounds, the role of NDT and the accuracy required in the inspection results in this process cannot be underestimated.

NDT Plan

- Define how to inspect, which components of the item, what methods, and what coverage.
- Invasive or noninvasive inspection?
- Consider feasibility and benefits.
- Need a structured decision process.
- Capability and effectiveness of the technique and reliability of results is key.
- Expected to be more proactive.
- Qualification and experience relevant to damage mechanism.
- Accuracy of results—need to check calibration and question the findings at the time of inspection.

Benefits

- Enhanced integrity through better understanding of potential deterioration mechanisms and vulnerable locations.
- More focused inspection strategies with optimized interval and inspection activities for each equipment item.
- Reduction in number of items inspected during each turnaround, through extended inspection intervals for items that are found to be overinspected.
- Reduction in unexpected deteriorations usually found during turnaround inspections which in turn leads to reduction in unplanned repairs, thus avoiding overrun of the shutdown period.
- Formally justified opportunities with recorded evidence to increase plant run-length time between turnarounds and optimize maintenance/inspection turnaround times.
- Better operational and maintenance control strategies to reduce equipment failure risks.

Inspection of Equipment

Before going for inspection, internal or external, of any equipment such as furnaces, columns, vessels, reactors, piping, etc., the inspector should update himself or herself with its complete history, like original thickness of the equipment or piping, corrosion allowances, corrosion rate, previous repair job undertaken, data on daily routine observations, and history of any plant upset or abnormal condition.

Furnace: The structural configurations of a furnace may be cylindrical, box, cabin, or multicell box. The radiant tube coil configurations include vertical, horizontal, helical, and arbor. The mode of firing may be upfired, downfired, or wallfired.

Visual Inspection

During visual internal inspection, the extent of external deposits and their colors should be observed. The condition of tube sheets and hangers and the welding integrity of the thermocouple points should be inspected. The hot spot locations, outer diameter (OD) growth, bulging, sagging, and tube displacement from its supports should also be observed.

External observations should include hot spot or corrosion of furnace box, the roll and plug leaks of the return headers, if provided, and the paint condition in general.

Detailed Internal Inspection

Internal inspection involves detailed examination of the in-service material degradation by measuring the following parameters mostly by NDT techniques and comparing with the design limits.

Scaling: Scaling denotes the metal loss. Color, magnetic nature, and measurement of scale thickness should be recorded.

Sagging/bowing: This results from unequal metal temperature distribution due to either flame impingement or coke formation inside the tube. All the tubes should be inspected for bowing. For tubes having welded return bends, actual measurements have to be taken with respect to some reference line to measure the extent of bowing. If the extent of sagging/bowing is more than 1.5 times the tube OD, the tube should be replaced.

Bulging: Bulging is caused by the loss of structural strength as a result of high metal temperature. For tubes having metal temperature in the creep range, one of the factors that should be taken into consideration in determining when to replace a tube is the amount of diametric creep that has occurred.

Bulging is measured by the OD gauge, set initially at 2.5% more than the OD. If the gauge gets stuck at any point, then OD at that point is accurately measured with a Vernier caliper or dial gauge OD caliper. It is common to limit the increase in diameter resulting from creep to 5% of the original external tube diameter.

Thickness survey: Ultrasonic thickness survey should be carried out along the length of all the tubes at a reasonable interval. In-roll and beyond-roll internal diameter (ID) should be measured wherever applicable for ascertaining the thickness loss. Thickness survey of bends is important, as these are prone to erosion also.

Welded joints: As welded joints are the regions of relatively high residual stresses, chances of in-service material failure/degradation is a common phenomenon, especially at the heat-affected zone (HAZ). All the accessible welded joints should be inspected by applying any or a combination of the following techniques:

Visual, magnetic particle, liquid dye penetrant, radiography, ultrasonic flaw detection technique.

Tube supports: Tube sheets and tube supports should be examined to determine their physical condition and fitness for further service. Tubes should be inspected at all the support locations for thinning and formation of grooves. Special attention should be given for wear/grooving etc., which may occur due to vibration at support guides in vertical furnaces.

Thermocouple Connections

The welding of thermocouple tips on the heater tubes often crack in service. All such connections should be checked.

Steam coils: Wherever steam coils are provided, the same should be inspected for corrosion. The steam coils should be hydrotested each turn around to check for any leaks.

Hydrotesting: Hydrotest of furnace tubes are carried at a test pressure of 1.5 times the maximum operable working pressure or as specified in the data card. Two numbers of gauges should be used—one at the highest point in the system and the other at the discharge of the hydraulic test pump. The test pressure is usually held on the heater for some minimum period of time, say 15–20 min, after leaks have been eliminated to a satisfactory degree.

Refractory lining: The refractory/insulation lining of the furnace should be inspected for general condition. Leakage of hot furnace gases through joints when the bricks or insulating concrete has fallen out exposes the supporting steel to high temperatures, rapid oxidation, and corrosion.

Ensure that expansion joints are fitted with asbestos rope/ceramic fibers. When localized fusion of the refractory lining is observed, a detailed study is required for finding its cause and deciding the corrective measures. If major refractory repairs are carried out in the furnace, the same should be properly dried during start-up.

Columns

A column is a cylindrical vessel placed in a vertical position. It is used as a heat and mass transfer equipment where fractionation of hydrocarbon components takes place based on boiling point difference and liquid–liquid extraction takes place based on density difference. Column internals usually consist of trays, structured and random packings, down comers, and demister pads.

External Inspection

Most of the external inspection can be done while the column is in operation. The following should be checked during external inspection:

Foundation and supports, foundation anchor bolts, vibration of ladders, stairways, platforms, structural etc., insulation and protective coatings, earthing connections, nozzles, and external inspection of metal surfaces. All the nozzles should be thickness surveyed from the outside.

Internal Inspection

Scaling: The scale samples should be collected and analyzed. Scaling represents metal loss.

Liquid level corrosion: Liquid level corrosion marks are often observed in the form of grooving on the shell at the liquid level. Even internals such as bubble caps can also be affected.

Distortion/displacement of internals: During plant upset condition, the column internals sometimes get displaced/distorted and thrown inside the column. Chimney trays, baffle trays, and wear plates may be disturbed from their positions. Channeling marks will be observed below the nozzle feed distribution header in case of blockage of a few perforations in the distribution header.

Corrosion and choking of demister pads: Demister pads are sometimes found corroded and choked. They should be inspected after steam cleaning.

Inspection of flash zone: The flash zone of any column should be thoroughly inspected as corrosion/welding damage of the impingement plate after long operation is common. The support beam above the flash zone should also be inspected for any distortion/corrosion.

Thickness survey from inside: Thickness survey of parent CS shell should be carried out where a corrosion-protective liner (SS cladding) is not present. For large-diameter CS nozzles, thickness survey can also be done from inside the column.

Pneumatic testing: Sleeved nozzles and reinforcement pads may be tested for their reliability.

Vessels

Pressure vessels are designed in various shapes. They may be cylindrical (with flat, conical, toriconical, torispherical, semi-ellipsoidal, or hemispherical heads), spheroidal, or spherical. Cylindrical vessels may be vertical or horizontal. Pressure vessels are used in most processes for the storage of volatile liquids.

External inspection: Skirts should be inspected for corrosion, distortion, and cracking from the outside as well as the inside. The weatherproofing on the extremities and fireproofing of structural supports should be checked for watertightness. The inside of the skirt is often subjected to corrosion. This is particularly true for vessels operating in a cryogenic environment.

Horizontal vessels resting on concrete saddle supports where water can accumulate and cause external corrosion should also be inspected. Horizontal vessels operating at high temperatures should be checked to ensure free thermal expansion. Foundation bolts should be inspected for corrosion and damage. The nuts on anchor bolts may be inspected to see that these are properly tightened. Crevice corrosion may exist around the heads and nuts of bolts. Ladders should be examined for free movement to take up expansion of the vessels.

Internal Inspection

Scaling: On vessel entry, first of all the scaling nature, color, and thickness should be ascertained. Excessive scaling is an indication of the corrosive nature of the fluid in service. The internal shell wall should be inspected for pitting after scales removal by wire washing.

Inspection of welded joints: The welded joints and HAZ should be checked visually for cracks. In case of doubt, they should be checked by a dye-penetrant test. The welded joints should be checked for leaks by pressurizing with air at a pressure of 1.03 kgcm^{-2} through the telltale hole provided in the reinforcing pad.

Inspection of nozzles: Nozzles inspection in a pressure vessel is very important, as these are relatively thin components with respect to the shell. Deteriorated nozzles should be replaced. New nozzles fabricated out of piping with a thickness equivalent to original nozzles are to be installed. Welding should be carried out from inside as well as outside with suitable electrodes matching the base metal and nozzle material.

Inspection of protective liner: Some pressure vessels are rubber-lined from the inside for protection against corrosion. The rubber lining should be inspected for mechanical damage, holes, cracking, blistering, bonding, etc. Holes in the lining are evidenced by bulging. A holiday detector should be used to thoroughly check the lining for leaks and holiday.

In case of a cladded vessel, the bulged, cracked, or heavily pitted cladding inside the pressure vessels should be repaired. The deteriorated cladding is to be removed by cutting. The edge of the remaining cladding is sealed by welding, using proper electrodes as per cladding and shell metallurgy. If the area of the damaged cladding is small, the area is weld overlaid using suitable electrodes. The area is then ground smooth. The repaired portion should be checked visually and by using D.P. for defects, cracks, etc.

Inspection of vessel wall: The vessel wall should be inspected for pitting. The pit depth should be measured and should not be more than one half of the vessel wall thickness exclusive of the corrosion allowance. The total area of the pits should not exceed 45 sq cm within any 20-cm diameter circle and the sum of their dimensions along any straight line within the circle should not exceed 5 cm.

Heat Exchangers

Prior to cleaning, all the accessible parts of heat exchangers should be inspected for fouling deposits, scaling, etc. If corrosion is observed beneath these deposits, then all such deposits should be analyzed. Tube bundles of stainless-steel material should be chemically treated with a solution of sodium carbonate or sodium nitrate before exposing them to atmospheric oxygen. For tube bundles where sulfide scales are likely to be available before opening, precautions should be taken to prevent overheating due to oxidation of pyrophoric iron.

Shell, Shell Covers, Channel, Channel Cover, and Floating Head Cover

General visual inspection of inside and outside surfaces and welds should be carried out for signs of pitting, grooving, scaling, erosion, or impingement attack. After visual inspection, a detailed thickness survey of these parts of the exchanger should be carried out using ultrasonic instruments.

Shell portions adjacent to the tube bundle impingement plate and baffle should be checked for erosion.

All the nozzles and small-bore connections should be checked for thinning.

The pass partition plates of the channel head and floating head cover should be inspected for corrosion, warping, and edge thinning.

Gasket surfaces should be visually checked for corrosion or mechanical damage. The sacrificial anodes, if required, should be replaced with new ones.

The lining, if provided, should be checked for any damage.

Tube Bundle

The tube sheets' gasket surface should be inspected for general corrosion, pitting, grooving, etc.

Tube-ends and pass-partition grooves in tube sheets should be checked for groove pits and erosion.

The peripheral tubes, tie rods, and baffles should be inspected for corrosion, erosion, cracks, etc. Nonferrous tubes like Naval Brass or Cupro Nickel should be checked for dezincification or denickelification. The clearance of peripheral tubes in baffle holes should be checked. If it is excessive, failure of tubes will occur due to vibration/erosion.

Hydrotesting should be done at the prescribed test pressures to ensure the integrity of the expansion joints and the tubes both under compressive and tensile forces (shell test and tube tests), respectively. Calibrated pressure gauges are to be used during the tests.

Sometimes tube leaks are detected during tube side pressurization. Again shell test is repeated for locating the leaky tube.

Demister Pad

The demister pad is to be inspected after opening, for chokage, corrosion, crimps, and oxidation due to high temperature. If the mesh size of the demister gets altered due to the above reasons, it should be replaced suitably with similar metallurgy demisters. During shutdowns the demister pads should be cleaned by utility steam.

Refractory

Refractory are nonmetallic materials suitable for construction or lining of furnaces operated at high temperature. In other words, nonmetallic materials that show no sign of incipient fusion when heated up to a minimum temperature of 15,000°C can be classified as refractory material. Refractory are not only used for temperature insulation but also as erosion/corrosion liner. Classification based on chemical composition, physical characteristics, and service application will result in the following type of refractory:

Silica, alumina-silicate, basic refractory, insulating fire bricks, abrasion resistance lining, special refractory.

Refractory are subject to failures depending on the following factors:

- Incorrect design/operating parameters and selection of refractory
- Thermal shock/explosions
- Vibration in heater tubes

- Mechanical failures
- Operating parameters (temperature)/process upset
- Incorrect use of refractory during repairs due to nonavailability of the correct refractory
- Improper workmanship
- Improper expansion joints—filler materials not correctly used
- Erosion

General inspection guidelines are useful:

- Look for detachment of refractory at the expansion joints.
- Look for the zones where there is a temperature gradient.
- Also look for the areas where there is change in shape of the equipment from the previous geometry.

In the selection of refractory lining, the following considerations are important:

Temperature and service condition of the equipment, various refractory properties like bulk density, cold crushing strength, thermal conductivity, chemical composition, RUL value (refractory under load), pyrometric cone equivalent, creep, apparent porosity, spalling resistance, etc.

For better performance, proper refractory application including the curing time is very important but is usually neglected during repair work.

There are separate procedures available for repairing insulating refractory and abrasion-resistant refractory. These should be repaired strictly as per the procedures given by the manufacturer/licensor.

Pressure Safety Valves

The following inspection and tests should be carried out:

1. Inspection and test of the valve in as-received condition. This is important and helps in establishing the frequency of inspection.
2. Visual inspection of different parts of the safety valve should be done after dismantling to check the following:
   a. Conditions of flanges for pitting, roughening, decrease in width of seating surface etc.
   b. Spring, for evidence of bending, corrosion or cracking, free length of spring.
   c. Bellows (if applicable) for damage.

   d. Position of setscrews and opening in bonnet.

   e. Inlet/outlet nozzles for evidence of deposits of foreign material and corrosion.

   f. Condition of external surface and evidence of mechanical damage.

   g. Body wall thickness.

   h. Conditions of steam, guide disc, nozzles, etc. for evidence of wear and corrosion.

   i. Seating surface of disc and nozzle should be critically examined for roughness or damage, which could result in valve leakage and may need correction. Care must be taken to ensure flatness of seats.

3. Pneumatic testing of the pressure-relief valve is carried out on the workshop test bench both in the as-received condition and after overhauling. The set pressure and tightness of the valve is witnessed by the inspection engineer and recorded in a register.

Inspection during Fabrication of New Piping

The inspection should include

- Study of tender document and all the technical specifications.
- Identification and inspection of material.
- Approval of welding procedures in accordance with code and tender requirement.
- Carrying out of performance qualification test.
- Ensuring that welding is carried out as per agreed procedures, by approved welders with specified electrodes.
- Inspections of welded joint fittings.
- Dye-penetrant examinations of the prepared edges for low-alloy steel and stainless steel.
- Ensuring proper preheating, maintaining proper interpass temperature, and post-welding heat treatment as specified.
- Radiographic and/or ultrasonic inspections of welded joints as specified.
- Ensuring repairs of the defective welds, if any, before giving clearance for hydrostatic testing.

- Ensuring proper repairs to damaged lining (cement/rubber), if any.
- Hydrostatic testing.
- Ensuring all approved deviations from the drawing are noted and as-built drawing prepared.
- Ensuring proper surface preparation and painting.
- Ensuring installation of proper insulation wherever applicable.
- Ensuring underground protection for buried piping has been provided as specified.

External Corrosion

The following areas of piping are prone to external corrosion.

- Piping above ground is subject to atmospheric corrosion.
- Pipelines touching the ground are subject to corrosion due to dampness of the soil crevice. Corrosion may take place at the pipe supports or sleepers where pipes are resting on them.
- Deterioration takes place at the pipe support locations where relative movement between pipe and pipe supports takes place.
- Buried pipelines are subject to soil corrosion.
- Underground pipelines are subject to corrosion due to the presence of stray currents.
- Impingement attack may take place on pipelines in the vicinity of leaky pipes and steam traps.
- Insulated lines where weather shielding is damaged are subject to external corrosion.
- Austenitic stainless-steel lines where chlorides can leach from external thermal insulation due to water are subject to stress corrosion cracking.
- Externally concrete-lined pipelines are subject to localized corrosion due to cracks in the concrete.
- Piping entering into or emerging from the underground may experience severe corrosion due to coating damage.
- Piping corrodes at locations of water accumulation and acid vapor condensation such as in the vicinity of fire hydrants, sulfur recovery plants, cooling towers, jetties, etc.

Approaches for Other Equipment Types

The fixed-equipment program described earlier focuses on only a part of a process plant's equipment. Fixed equipment, along with safety instrumented systems, are the items that generally have the highest risk and that are most directly addressed by regulatory requirements. However, the reliability of rotating equipment, electrical power distribution, other instrumentation and controls, or functional aspects other than containment of some fixed equipment also have substantial impact on safety and business needs. Preventive maintenance provides the basic care of equipment including predictive or condition-monitoring tasks. The term "preventive maintenance" as commonly used has different meanings to different people. For some, it includes predictive and condition-monitoring activities; to others, those are a separate set of activities. Preventive maintenance is defined as "maintenance carried out at predetermined intervals or according to prescribed criteria and intended to reduce the probability of failure or the degradation of the functioning of an item" per API 689. Preventive maintenance consists of two major types of activities.

Scheduled Restorative or Discard Tasks

- This is renewing a condition, consumable, or a wearing part.
- Examples include oil changes, greasing, filter changes, cleanings, annual overhauls, quarterly calibration, weekly inspections, other wearing part changes, or adjustments on a scheduled basis.
- This subpart alone is sometime called preventive maintenance.

Predictive Maintenance or Condition Monitoring

- An equipment maintenance strategy based on measuring the condition of the equipment in order to address whether it will fail during some future period, and taking appropriate action to avoid consequences of that failure.

10

Shutdown Safety

During shutdown execution the management must emphasize on achieving the safety goal. It requires a coordinated effort by all stakeholders involved in the shutdown process. While excellent performance is expected throughout plant operations on a day-to-day basis by own employees, additional effort and rigor during shutdowns is required due to increased activity levels and involvement of many contract workers. The shutdown manager must assure that potential physical, mechanical, chemical, and health hazards are recognized and provisions are made for safe operating practices and appropriate protective measures are in place. These measures may include hard hats, safety glasses and goggles, safety shoes, hearing protection, respiratory protection, and protective clothing such as fire-resistant clothing where required. In addition, procedures should be established to assure compliance with applicable regulations and standards.

The shutdown team needs to have experience in plant operations, hazardous materials management (asbestos, lead, polychlorinated biphenyls, mercury, catalyst handling), occupational hygiene, and construction safety to ensure success and prevent incidents from occurring. A pre-assessment of work areas should be completed by the shutdown team during the pre-planning stage of the shutdown. The assessment should involve a discussion of work scope and a visual assessment of the work area for potential worker exposure risks. Potential risks need to be flagged and brought to the attention of the shutdown team. A detailed plan then needs to be developed for dealing with the hazards. Occupational health and safety hazards that need to be integrated into the shutdown plan include airborne contaminants (asbestos, fly ash, coal dust, catalyst dusts, welding fumes, refractory ceramic fibers, lead, radiography); personal fall protection:

- Falling objects, eye protection, slipping and tripping hazards, proper storage of gas/air cylinders
- Confined space entry and vessel ventilation

One of the major hazards during shutdown is exposure to hazardous substances, as these are a deviation from routine operations. Plant turnarounds require careful planning, scheduling, and step-by-step procedures to make sure that unanticipated exposures do not occur. Any plant shutdown requires a complete plan in writing to cover all activities, the impact on other operations, and emergency planning. Plans are normally formulated by plant personnel in conjunction with contractors.

Hazardous Chemicals and Catalysts

In a refinery, hazardous chemicals can come from many sources and in many forms. In crude oil, there are impurities such as sulfur, vanadium, and arsenic compounds. These components are inherently hazardous to humans, as are the other chemicals added during processing. The main hazards in this process come from possible exposure to the catalysts, hydrofluoric acid or sulfuric acid, and their dusts, by-products, and residues, as well as hydrogen sulfide, carbon monoxide, heat, and noise. Other processes utilize acid catalysts and caustic "washes." These can lead to hazardous situations, especially in shutdowns where a contractor's personnel may be exposed to residues or other contaminants. Other hazards include fire, explosion, toxicity, corrosiveness, and asphyxiation. Information on hazardous materials manufactured or stored in a refinery should be supplied by the client's representative when a work permit is issued. Information is required from refinery personnel and specialized training is required in the necessary procedures and personal protective equipment, including its care and use.

Health and Hygiene Hazards

Table 10.1 reviews common hazardous chemicals and chemical groups typically present and their most significant hazards to workers. Care should be exercised at all times to avoid inhaling solvent vapors, toxic gases, and other

TABLE 10.1

Common Hazardous Chemicals

Material	Dominant Hazard
Additives	Usually skin irritants
Ammonia	Toxic on inhalation
Asbestos	Designated substance under construction
Carbon monoxide	Toxic on inhalation
Caustic soda	Corrosive to skin and eyes
Chlorine	Corrosive to skin and tissue on contact or inhalation
Hydrogen sulfide	Toxic on inhalation
Nitrogen	Asphyxiant
Silica	Designated substance under industrial regulations
Sulfuric acid	Corrosive to skin and tissue on contact or inhalation
Sulfur dioxide	Toxic on inhalation

respiratory contaminants. Because of the many hazards from burns and skin contact, most plants require that you wear long-sleeved shirts or coveralls.

Common Hazards

Fire and Explosion

Other principal hazards at refineries are fire and explosion. Refineries process a multitude of products with low flash points. Although systems and operating practices are designed to prevent such catastrophes, they can occur. Constant monitoring is therefore required. Safeguards include warning systems, emergency procedures, and permit systems for any kind of hot or other potentially dangerous work. These requirements must be understood and followed by all workers. The use of matches, lighters, cigarettes, and other smoking material is generally banned in the plant except in specially designated areas.

Confined Spaces

On most jobsites, there are potential confined space hazards. These hazards are multiplied, however, on a refinery site because of the complex collection of tanks, reactors, vessels, and ducts combined with a wide variety of hazardous chemicals and emissions, often in enclosed areas. Many of these chemicals can produce oxygen-deficient, toxic, or explosive atmospheres. Knowledge of general confined space procedures and specific in-plant requirements is critical in refinery work.

Confined Space Entry and Vessel Ventilation

Confined space work during shutdowns can be associated with refractory removal and replacement, chemical cleaning, catalyst handling, welding/ gouging, and coating applications to name a few. In these "closed-in" environments, there is usually an increased potential for worker exposure. Attention needs to be given to heat stress, personal protective equipment and ventilation requirements. All personnel entering into a confined space must have specialized training.

Vessel ventilation is critical prior to workers entering into the confined space. The key reasons vessels are ventilated are to purge the vessels of process contaminants such as hydrocarbons, inert gases, and steam; for comfort ventilation; for heat stress control; and to control generated contaminants. The success of ventilation is verified by gas testing prior to entry. Gas testing typically includes testing for lower explosive levels, oxygen percentage, toxics such as hydrogen sulfide (H_2S), and carbon monoxide (CO).

Airborne Contaminants

In dealing with airborne contaminants, the shutdown team's goals are to control risk of worker exposure to airborne fiber; to control risk of airborne fiber release into the plant environment; and to identify improvements to programs.

Personal Fall Protection

Workers exposed to falls need to meet the minimum requirements of provincial/local occupational health and safety (OH&S) regulations. If handrails or floor gratings are removed to facilitate work, workers must use approved fall restraint or fall arrest equipment. Other workers must be protected by either temporary scaffold guardrails or a flagged/ribboned area a minimum of 6 ft from the fall hazard. The flagged area should not be left unattended for long periods of time.

Falling Objects

Job tasks that pose a risk of falling objects should be flagged off. A plant procedure should be developed and understood for flagging and ribboning. Past experiences show near-miss incidents occur with falling objects include hoisting materials and equipment; storage of tools and material laid down inside the toe boarded area of a scaffold or work platform; passing of materials or equipment from hand to hand outside the bounds of the handrails and toe boards, work platforms, and scaffolds.

Eye Protection

Safety goggles are required during work tasks that involve grinding, buffing, and all cut and saw operations.

Slip and Trip Hazards

Slipping and tripping hazards are common during shutdowns, and can be minimized by ensuring contractors implement proper housekeeping procedures. Wet materials such as oil, wet ash, and water should be cleaned up immediately or flagged off as slip hazards.

Areas used for lay-down equipment must be arranged to prevent tripping hazards in common walkways. Hoses, cords, and air lines must be arranged to prevent tripping hazards in walkways. Scaffold support piping and hoist support piping (typically used for hoists at floor levels) should be flagged and ribboned for visibility.

Proper Storage of Gas/Air Cylinders

All air and gas cylinders should have the main valves closed before workers leave their areas for all breaks. Regulators should be removed and safety caps installed at the end of each work shift. All cylinders should be securely stored at all times.

Safety Planning

The participation of owners in safety starts at the very beginning and lasts till the end. Owner involvement in safety planning includes selecting safe contractors, addressing safety in design, including safety requirement in the execution, and being actively involved in the shutdown safety management.

Owners must make a group responsible for planning, developing, organizing, implementing, and auditing of a safety system for the shutdown job. They must develop a structure to assure implementation of the commitment to safety and health at work by own employees and contract workers. Owners must ensure

- In-house safety rules are in place to provide instructions for achieving safety management objectives.
- Training is imparted to equip personnel with knowledge to work safely and without risk to health.
- Evaluation of job-related hazards or potential hazards and development of safety procedures are done well ahead of execution.
- Promotion of safety and health awareness at the workplace.
- Evaluation, selection, and control of contractors and subcontractors to ensure that they are fully aware of their safety obligations and are in fact meeting them.
- Emergency preparedness to develop, communicate, and execute plans prescribing the effective management of emergency situations.

Safety Preplanning

An important aspect of preplanning is to integrate occupational health and safety into the overall shutdown plan. Ownership needs to be created by the shutdown team, which will involve key stakeholders, SHE professionals, and contractors. Unique hazards occur during shutdowns that sometimes are not well understood by plant personnel. It is therefore important during the preplanning stages that a clear work scope is prepared and a plan is put into place to manage all safety and occupational hazards.

To bring increased focus on occupational health and safety during shutdowns, it is required to document safe work plans and hazard assessments for all work scope. A list of safety responsibilities for managing safety and health by everyone involved in the work activity should be followed. A list of risks and the risk assessment results, including the contingency plans for foreseeable emergencies should be noted. The procedures for monitoring and assessing the implementation of these actions should be carried out. Occupational health and safety budgets and resources need to be considered for collection and analysis of hazards such as asbestos, lead, refractory, silica, welding, and off-gassing materials. Specialized instruments such as photoionization detectors, H_2S detectors, are excellent for quick detection. But sometimes are expensive and not easy to obtain.

Evaluation and Selection of Contractors and Subcontractors

Evaluation and Selection Strategy

The evaluation and selection strategy should clearly aim at ensuring that contractors and subcontractors with knowledge of good safety standards and a good record of putting them into practice would be selected for the work. A practicable approach to evaluate and select suitable contractors and subcontractors is set out below:

1. Each contractor and subcontractor wishing to qualify as a bidder should be asked to provide a safety policy, which would be vetted to assess its adequacy.
2. The contractor and subcontractor should also be required to submit the safety organization and details of responsibility, track records in safety, working experience with clients of high safety standards, safe systems of work/safety programs in place, current safety management system, and training standards. These should also be vetted to assess adequacy.

Pre-bid briefing should be provided to all qualified bidders. They should be invited in writing to attend the briefing and their attendance should be recorded. Provisions in the contract, as well as the local safety laws that apply, should be clearly communicated to bidders in the briefing.

Bidders should identify all of the safety and health hazards within the specifications. To help them do this, a checklist on all the common safety and health problems, which may arise during the work should be provided/communicated to them in the specifications before the bid is made.

- The bids submitted by the potential subcontractors should be checked against the potential safety and health problems to ensure that all the safety and health hazards that may arise during the work have been clearly identified by them and that proper provisions have been made for the control of the risks assessed. Each potential subcontractor should also be required to submit an outline safety plan for the implementation of the risk control measures. The outline safety plan should be set out in summary from the subcontractor's proposed means of complying with his or her obligations in relation to safety and health at work.

- The contract should go to the subcontractor who is able to identify all the safety and health hazards that may arise during the work, can assure that the most proper and adequate provisions have been made for the control of the risks, and has the best outline safety plan compared to other bidders.

The contractor should provide all safety and personal protective equipment (PPE) required to complete the contracted scope of work.

The contractor should ensure that the supervisor on site is well trained on environment, health, and safety activities and regulations in the performance of the work.

Safe Work Practices and Procedures

Site Access and Use

All sites must have controlled access to limit unauthorized individuals from entering the construction area.

Cleanup

The contractor shall keep the work area, specifically walking and working surfaces, clean and free from debris and trash, which could cause slipping and tripping hazards. Tools, materials, dirt, lumber, concrete, metal, insulation, paper, etc. should be promptly cleared and disposed of by the contractor. All debris should be disposed of each day off the campus or in a contractor-supplied dumpster.

General Safety

- Hearing protection and safety glasses must be worn in all operating areas or as posted.
- Respiratory protection or equipment must be fit-tested. Facial hair is unacceptable where the mask must make an airtight seal against the face.
- Shirts must be long-sleeved and worn with full-length pants or coveralls.
- Clothing must not be of a flammable type such as nylon, Dacron, acrylic, or blends. Fire-resistant types include cotton, Nomex, and Proban.
- Other PPE required may include acid hood, impervious outerwear, rubber boots, face shields, rubber gloves, disposable coveralls, monogoggles, and fall-arrest equipment.

- Smoking is allowed only in designated areas.
- Vehicle entry is by permit only.
- Vehicles must be shut down at the sound of any emergency alarm.

Evaluation of Job-Related Hazards

This refers to carrying out hazard analysis for jobs. The objective of hazard analysis should be to provide a means whereby job hazards or potential hazards are identified, evaluated, and managed in a way that eliminates or reduces them to an acceptable level. Safe working procedures and safety precautions that are to be taken to prevent the hazards and to control the risks should be developed after the hazard analysis.

Hazard identification is the process of identifying all situations or events that could give rise to the potential for injury, illness, or damage to plant or property. Hazard identification should take into account how things are being done, where they are done, and who is doing them, and should also consider how many people are exposed to each hazard identified and for how long.

The following should be accorded top priority in the hazard identification process:

- High frequency of accidents or near misses
  Jobs or works with a high frequency of accidents or near misses pose a significant threat to the safety and health of employees at work.

- History of serious accidents causing fatalities
  Jobs or works that have already produced fatalities, disabling injuries or illnesses, regardless of the frequency, should have a high priority in the hazard identification process.

- Existence of a potential for serious harm
  Jobs or works that have the potential to cause serious injury or harm should be analyzed by the hazard identification process, even if they have never produced an injury or illness.

- Introduction of new jobs or work
  Whenever a new job or work is introduced, a hazard identification process should be conducted before any employee is assigned to it.

- Recent changes in procedures, standards, or legislation
  Jobs or work that have undergone a change in procedure, equipment, or materials, and jobs or work whose operation may have been affected by new regulations or standards should require the carrying out of hazard identification process.

PPE

In the hierarchy of control measures, PPE should always be regarded as the "last resort" to protect against risks to safety and health. Engineering controls and safe systems of work should always be considered first. It may be possible to do the job by another method, which will not require the use of PPE or, if that is not possible, to adopt other more effective safeguards.

However, in some circumstances, PPE will still be needed to control the risk adequately.

PPE includes the following, when they are worn for protection of safety and health:

1. Protective clothing such as aprons, protective clothing for adverse weather conditions, gloves, safety footwear, safety helmets, high-visibility waistcoats, etc.

2. Protective equipment such as eye protectors, hearing protectors, life jackets, respirators, breathing apparatus including those used underwater, and safety harness

The shutdown manager and the contractors should ensure that appropriate PPE and training for its usage is provided to the workers.

1. *Conducting PPE risk assessment*: If it is necessary to provide PPE, an enterprise should conduct an assessment. The purpose of the assessment is to ensure the correct PPE is chosen for the particular risk. Except in the simplest and most obvious cases that can be repeated and explained at any time, the assessment will need to be recorded and kept readily accessible by those who need to know the results.

2. *Proper selection of PPE*: The proprietor or contractor of an enterprise should determine what type of PPE is required, taking into consideration the legal requirements for specific situations, the intended use of the PPE, the manufacturer's product standards, the ergonomics of the design of the PPE, acceptability of the PPE to its wearer and user, and, if used in conjunction with other PPE, compatibility with the PPE in question, etc.

3. *Steps to ensure adequate supply of PPE*, including replacement supply and spare parts.

4. *Steps to ensure adequate training, information, and instruction* to workers on the safe and proper use and maintenance of PPE.

Permit Requirement

No work takes place in a refinery and process plant without a safe work permit. A safe work permit is a document issued by an authorized representative of the client permitting specific work for a specific time in a specific area. Work permits should indicate the date and time of issue, the time of expiry, a description of the work to be done, and the name of the company performing the work. Permits also specify any possible hazards and any protective equipment needed for the job. The permit will advise you on any steps required to make the area or equipment safe for work, tell you the results of any gas tests, advise you on any electrical lockouts that have been done, and tell you about any work practices required for the specific job. Safe work permits are valid only for a limited time and must be renewed following expiry. Normally during a shutdown, short-term permits are issued rather than ongoing permits. This means the contractors and line supervisors must check daily or more frequently to ensure that hazardous activities are coordinated well and remedial actions are taken. After an emergency event, any required gas testing or other testing must be repeated to ensure a safe return to the work.

Owners must make a group responsible for planning, developing, organizing, implementing, and auditing of a safety system for the shutdown job. They must develop a structure to assure implementation of the commitment to safety and health at work by own employees and contract workers.

The types of safe work permits required typically include the following; specific categories may vary from site to site:

- Hot work—covers any work that involves heat or an ignition source, including welding, grinding, and the use of any kind of motor. In high-risk areas, a spark watch may be required.
- X-ray and radiation.
- Confined space entry hot work—involving potential ignition hazards.
- Confined space entry cold work—involving work that will not produce a spark.
- Hoisting—permit.
- Electrical—for other than routine work.
- Camera—typically requires a hot work permit when lighting is required.
- Vehicle movement.
- Excavation.

Training

Oil refinery workplace safety also involves extensive training of workers in safety standards. Workers are educated on processes that are designed to prevent potential disasters. Creating a safe workplace through regulations and safety procedures can reduce accidents and injuries.

Executing

1. Provide adequate and appropriate resources to implement safety.
2. Place occupational safety and health as one of the prime responsibilities of line management, from the most senior executive to the first-line supervisory level.

3. Ensure that employees at all levels receive appropriate training and are competent to carry out their duties and responsibilities.

4. Provide adequate and effective supervision to ensure that the policies and the plans are effectively implemented.

5. Determine and execute plans to control the risks identified.

6. Arrange safety audits and periodic status analysis as an independent check to the efficiency, effectiveness, and reliability of the safety management system, and carry out the required corrective actions.

7. Motivate all employees by a combination of rewards and sanctions and stress on the reinforcement of the positive behavior contributing to risk control and the promotion of a positive safety culture.

Facilities

The following site facilities are required during the execution phase.

Washing facilities: Adequate and suitable facilities for washing should be provided for the use of the persons employed for the construction work.

Toilets: Suitable and sufficient conveniences are to be provided for the exclusive use of the males engaged or employed in or around the place of work, and for the exclusive use of the females employed in or around the place of work.
Where there are more than 15 employees, the minimum requirements are

- One urinal where 15 males are employed, plus one urinal for each 30 males or part thereof
- One water closet where 15 or less employees are employed, plus closets for each of the 30 employees or part thereof

Where females are engaged or employed, there should be suitable provision for the disposal of sanitary towels.

Drinking water: An adequate supply of wholesome drinking water should be provided to all persons on the construction work in accordance with the following conditions:

- It should be readily accessible to all persons engaged in the work and clearly labeled as drinking water.

First aid facilities: Every employer is required to provide adequate first aid facilities, appliances, and requisites. Normally, first aid is to prevent the condition of the injured person from becoming worse until more skilled help becomes available.

Lighting: Poor light can be a major contributor to accidents, and employers must ensure that adequate lighting is provided. Lighting should be provided over the entire place of work.

Housekeeping (general): Confusion will be reduced and operations will be more efficient when the work area is orderly and tidy. Tidiness and safety go hand in hand. Untidiness causes accidents—for example, employees trip over objects, slide on greasy surfaces, cut hands on shutdowning nails, or walk into poorly stacked materials, or vehicles run over or back into materials, plant, or employees.

All rubbish around machinery, plant, accessways, stairwells, site facilities, and the site in general should be regularly gathered and disposed of in suitable bins.

Where rubbish is burned on site, this should be done in a suitable incinerator well away from flammable material. Appropriate permits may need to be obtained to burn rubbish on site.

Safe access: Employers should ensure that

- All employees are able to reach their place of work safely.
- There are safe roads, gangways, passageways, ladders, and scaffolds.
- All walkways are level and free from obstructions.
- Materials are stored safely.
- Holes are securely fenced or covered and clearly marked.
- Chutes are available to avoid waste being thrown down in the open.
- Nails in reused timber are removed or hammered down.
- There is adequate lighting provided for the work
- Temporary props and shoring are in place to support temporary work as necessary

Working at Heights over 3 m: The first and essential step in ensuring that work is done safely is to ensure that it is practicable for the work to be carried out safely. Safe work practices may include one or more of the following:

- Guarding
- Safety nets
- Fall arrest systems

Guardrails/toeboards: A toeboard must be installed where there is a risk of tools or materials falling from the roof/place of work.

Safety nets: Safety nets can provide a satisfactory means of protection against falling, while allowing workers maximum flexibility of movement.

Fall-arrest systems: Individual fall-arrest systems include inertia reel systems, safety harnesses, lanyards, and static lines. People required to use this equipment must be trained in its use.

Scaffolding: It is important that a competent person inspect all scaffolding:

- Before it is used
- At least weekly while it is in use

- After bad weather or any other occurrence that could affect its stability
- After periods where the scaffold has not been used for some time

The person carrying out such inspections should ensure the following requirements are met:

- There is proper access to the scaffold platform.
- All uprights are properly founded and provided with base plates. Where necessary, there should be timber sole plates, or some other means used to prevent slipping and/or sinking.
- The scaffold is adequately braced to ensure stability.
- Load-bearing fittings are used where required.
- Uprights, ledgers, braces, or struts have not been removed.
- Working platforms are fully planked, with the planks free from obvious defects such as knots and arranged to avoid tipping and tripping.
- All planks are securely restrained against movement.
- There are adequate guardrails and toeboards at every side from which a person could fall.

Electrical supply: Electricity is almost universally used on construction sites as a power source for a range of machinery and portable tools. In addition, lighting and heating are in wide use on construction sites.

Temporary supply switchboards: All supply switchboards used on building and construction sites should be of substantial construction and should

- Where installed in outdoor locations, be so constructed that safe operation is not impaired by the weather
- Incorporate a stand for the support of cables and flexible extension cords
- Be provided with a door and locking facility acceptable to the electrical supply authority

Safe use of chemicals: Many different chemicals and hazardous substances are used in construction work. It is essential that effective control measures and policies are developed and implemented to ensure that the chemicals or substances are used safely.

These measures should include policy on

- Safe handling of chemicals or hazardous substances
- Correct storage procedures to be adopted
- Safe transportation procedures
- Safe disposal procedures

The safe handling of chemicals should be aimed at eliminating or minimizing risk to employees or others, and will involve reading labels and material safety data sheets (MSDS) and complying with instructions.

All chemicals should be stored in their original containers in a safe, well-ventilated, secure place and in accordance with directions on their labels and MSDS.

If chemicals are transported, regulatory requirements should be observed in respect to documentation, compatibilities, and load security.

All employees should be trained in the correct practices to be followed when using chemicals and hazardous substances, and in how to deal with emergencies that may arise while using any substance.

Cranes and lifting appliances: Cranes and other lifting appliances are valuable assets on construction sites. It is important, however, to pay regular attention to certain aspects of their operation. Employers and employees using cranes and lifting appliances should ensure that

- Cranes are inspected weekly, and thoroughly examined every 12 months by a competent person. The results of inspections should be recorded.
- There is a current test certificate.
- The driver is trained, competent, and over 18 years of age.
- The controls (levers, handles, switches, etc.) are clearly marked.
- The driver and dogman determine the weight of every load before lifting.
- Every jib crane with a capacity of more than 1 ton has an efficient automatic safe load indicator that is inspected weekly.
- A hydraulic excavator being used as a crane has the maximum safe load clearly marked and hydraulic check valves fitted where required.
- The crane is on a firm level base.
- There is enough space for safe operation.
- The dogman has been trained to give signals and to attach loads correctly and knows the lifting limitations of the crane.
- If it can vary its operating radius, the crane is clearly marked with its safe working loads and corresponding radii.
- The crane is regularly and thoroughly maintained.
- The lifting gear is in good condition and has been thoroughly examined.

Confined spaces: Confined spaces are not limited to closed tanks with restricted means of entry and exit.

Also included are open manholes, trenches, pipes, flues, ducts, ceiling voids, enclosed rooms such as basements, and other places where there is inadequate ventilation and/or the air is either contaminated or oxygen-deficient.

Before entry into any confined space, it should be tested to determine that there are adequate levels of oxygen present, and that dangerous amounts of flammable and/or poisonous gases are not present.

No one is to enter any space if testing shows that the air is dangerous inside. Forced ventilation should be used to remove or dilute the gases and supply fresh air. The air should be tested again prior to entering, and monitoring continued while work is being conducted inside the space.

Among the confined spaces that may be inherently hazardous are

- Manholes, tunnels, trenches set in chalk soil, which can partly fill with carbon dioxide gas, displacing breathable air
- Poisonous or flammable gases can collect in manholes in contaminated ground (e.g., near underground petrol tanks or refuse tips)
- In manholes, pits, or trenches connected to sewers, there can be a buildup of flammable and/or poisonous gases and/or insufficient oxygen in the air
- Sludges and other residues in tanks or pits, if disturbed, may partially fill the confined space with dangerous gases

Precautions

If work in a confined space could be potentially dangerous, entry should be strictly controlled and detailed precautions taken.

Preferably, employers should adopt an entry permit system, so as to ensure that employees and others are aware of the location of anyone required to enter confined spaces. As mentioned earlier, tests may be required to identify any dangerous amounts of flammable or poisonous gases.

Where the confined space itself may be dangerous (regardless of the work carried out):

- People who are required to work in or enter the space should receive training and instruction in the precautions to be taken inside the area.
- At least one person should be stationed outside the space to keep watch and communicate with anyone inside.
- Rescue harnesses should be worn by all those inside the confined space, with a lifeline attached to the harness and a suitable winching mechanism at or near the point of entry.
- Rescue procedures should be included in the training of workers. Reliance should never be placed on one person alone to lift injured or unconscious people out of a confined space during rescue, unless they are equipped with special lifting appliances. Rescue equipment, including emergency breathing apparatus, should be available near the entrance at all times.

- No attempt should ever be made to clear fumes or gases with pure oxygen.
- Appropriate respiratory protection shall be provided where the results of monitoring assessment indicate that a safe atmosphere cannot be established.

Electrical Precautions

- Electrical tagging and lockout procedures must be understood and followed by all workers.
- All electric tools, cords, and equipment must be grounded or double-insulated.
- Use explosion-proof fixtures where required.

Sewers

- Sewers must be covered when hot work is being done in the vicinity.
- Sewer covers must be in good condition with no openings for vapor flow.
- Sewer covers are to be removed when hot work is discontinued at the end of the job or overnight to accommodate drainage.

Blinding or Blanking-Off

- Piping connected to a work area from vessels, pumps, and other sources is isolated or blinded with a solid plate prior to the start of work.
- Blanking can sometimes be done with two valves and a bleeder valve between them. In this case, the valves should be closed, chained, locked, and tagged.

Control Strategy

The control strategy should aim at monitoring the safety performance of the subcontractor and keeping him or her on the right track with regard to the achievement of the client's safety and health objectives during the execution of the contract. A practicable control approach should include the following:

1. *Special terms and conditions in the contract*: All safety rules and provisions should be set down in detail in the contract for the contractor to follow and implement. One of these provisions should be that the subcontractor agrees to abide by all the provisions of the client's safety policy, which may affect his or her employees' work,

including compliance with workplace safety rules. In case the contractor further subcontracts all or part of his or her work to other subcontractors, the contractor should ensure that the subcontractors are fully aware of the safety policy and safety rules of the owner.

The following special conditions should therefore be attached to the contract for the contractor to undertake: to inform any subcontractor of all safety requirements; to incorporate observance of all safety requirements as a requirement of any future subcontract; to require the sub-subcontractor to do similarly if he or she in turn subcontracts any work. Another provision in the contract should be that the subcontractor is required to submit a detailed and comprehensive safety plan based on the outline safety plan, indicating how he or she and the subcontractors (if any) are going to implement the safety measures for risk control during the work. The safety plan should include detailed policies, procedures, rules, safety obligations, and responsibilities of the contractor which, when being implemented, should ensure compliance with all safety rules set out in detail in the contract. The subcontractor should adhere to the safety plan in carrying out his or her obligations under the contract and should ensure that his or her own subcontractors of any tier (if any) receive copies of the safety plan and comply with its requirements as well.

2. *Risk assessment by the subcontractor before the commencement of the work*: The subcontractor should be requested to conduct a risk assessment before the work commences and recommend a safe system of work for the work. The safe system of work should include safety precautions, work methods, tools and equipment to be used, and how the subcontractor organizes his work to reduce risks to employees' safety and health. The subcontractor will be required to submit the risk assessment report together with the organization chart and expected workforce, list of the subcontractors and their key personnel, safe working procedures, list of tools and equipment, and the preventive maintenance schedule, etc.

3. *Control of subcontractor on site*: The following are some of the approaches for controlling the safety performance of a subcontractor on site. The contractor will be required to appoint/nominate a person or a team to coordinate all aspects of the contract, including safety and health matters on site. In addition, the contractor should develop communication paths to pass on all relevant safety information to those in the shop floor level. The contractor will be required to attend a pre-contract commencement meeting with the client to review all safety aspects of the work. The proprietor or contractor should carry out regular safety inspections daily to check on a contractor's activities. The contractor will be required to provide written method statements in advance of undertaking any special work such as demolition, confined space

work, working with asbestos, work on energized electrical installations, erection of false work or temporary support structures, erection of steel, and any work that involves disruption or alteration to main services or other facilities of the client's activities. The contractor will be required to report to the client all lost time, accidents and dangerous occurrences, including those of subcontractors. All employees of the contractor should be properly and effectively trained in safety and health matters in connection with the job requirement.

Monitoring/Auditing

To carry out the proactive monitoring of performance by, for example, surveillance and inspections of both hardware (i.e., premises, plant, and substances) and software (i.e., people, procedures, and systems of work), and monitor the degree of compliance with the safety and health arrangements of the enterprise.

A safety auditor should be appointed to periodically conduct a safety audit.

Immediate causes of deficiency should be determined for the substandard performance and the underlying causes identified.

Necessary actions should be taken on the safety audit report submitted, including drawing up a plan for improvements to the safety management system and implementation of the plan.

How a safety system can be enforced properly

The written safety program is the backbone of any company safety policy. A formal written program covering the regulations that apply to your specific job should be framed. A compact field manual is also to be provided for distribution to the field personnel. The manual is designed in a notebook format to make it easy to update the program. It should include

1. Toolbox talks
2. Hazardous communication
3. Safety checklists
4. Fleet policy
5. Substance abuse policy
6. Emergency procedures
7. Respiratory protection
8. Disciplinary action plans

Promotion of Safety and Health Awareness in the Workplace

- Evaluation of job-related hazards or potential hazards and development of safety procedures.
- Training to equip personnel with knowledge to work safely.

- In-house safety rules to provide instruction for achieving safety management objectives.
- A program to identify hazardous exposure or the risk of such exposure to the workers and to provide suitable PPE as a last resort where engineering control methods are not feasible.

Safety Department

A safety department or similar setup should be established to coordinate the implementation of safety plans or programs by the line management. Its primary role is to advise the line management on safety and health practices, requirements, and standards. It should not play a "line" and certainly should not be held accountable for the consequences of the lack of control on the shop floor or at the site.

The safety office or safety department should have the following main roles:

1. To serve as a resource person and in-house safety consultant.
2. To plan and prepare safety programs.
3. To advise top management and line management on safety and health matters.
4. To coordinate the implementation of safety plans and programs.
5. To monitor compliance and implementation of safety plans and programs.
6. To track corrective actions and verify the effectiveness of safety matters.
7. To serve as a trainer in safety matters.

Line management (including managers and supervisors) has the following responsibilities.

- To assist in the implementation of policies and procedures.
- To assist in the identification, assessment, and elimination of hazards and the control of risks.
- To supervise employees to ensure safe and correct working procedures.
- To ensure that effective consultation on safety and health matters occurs.
- To investigate accidents and incidents at work.
- To participate in induction and ongoing safety training programs for employees.
- To respond to safety initiatives of safety adviser/safety officer, safety supervisor, or other employees and to the safety advice from government officers.

- To communicate effectively the hazards to employees and keep abreast of current safety and health legislation and information.
- To submit periodically to senior management statistics and reports concerning safety and health performance.

Safety adviser: A safety adviser should have the responsibility to assist the top management and senior management in promoting the safety and health of employees in the enterprise. His main duties should include the following:

- To identify and assess the hazards at work.
- To work with senior management or line management to eliminate or control these hazards by advising them regarding measures to be taken, and, with their endorsement, implement such measures.
- To resolve shop floor safety and health issues.
- To conduct safety and health inspections to check the safety performance and recommend corrective action to senior management or line management.
- To investigate industrial accidents and incidents and recommend remedial measures to prevent recurrence.
- To be well informed about workplace safety performance.
- To consult with senior management, line management, and employees about changes in the workplace which would likely affect the safety and health of employees at work.
- To report regularly to the top management and senior management about the safety and health performance in the enterprise.

Identifying training needs: To equip his or her workers with knowledge on work safety and health, the proprietor or contractor of an enterprise must first identify what their safety and health training needs are. These needs are best established as part of an overall training needs analysis. For existing jobs, he or she can do the following:

1. Consult job-specific accident, ill-health, and incident records to see what caused loss of control and how it can be prevented.
2. Gather information from employees about how the work is done.
3. Observe and question workers when they are working, to understand what they are doing and why. This may be particularly relevant in complex process plants where any analysis has to take account of all the possible consequences of human error.
4. Consult risk assessment for the work.

Emergency preparedness: Emergency preparedness is vital because, when an emergency does occur, a quick and correct response is necessary to reduce injuries, illnesses, property damage, environmental harm, and public concern.

Emergency planning: A list of potential emergency situations such as fire, electric shock, flood, explosion, hazardous chemical spills or releases, internal/external leaks of explosive or flammable gas, personal injuries and illnesses, natural disasters, electrical outage, and critical damages to facility/equipment should be drawn up, with priority.

Emergency Warning System and Procedures

In oil refineries, there will be both plant alarms or whistles and individual unit alarms. All workers must receive training in recognizing and responding to these alarms. Verbal messages usually accompany the alarms. There will be different alarms for a fire emergency and toxic emergency.

- When an alarm sounds, secure all equipment and shut down all vehicles.
- Note the wind direction (wind socks) and proceed to the appropriate assembly area (or safe haven).
- Do a head count to make sure all personnel are accounted for and report the result to a client contact person.
- Know the local designated safety areas or safe havens and emergency phone number(s).
- Proceed to the assembly area.

Safety Promotion Program

The objective of safety promotion is to develop and maintain awareness among all personnel, of the organization's commitment to safety and health, and of the individual employee's responsibility to support that commitment. Safety promotion programs should be developed and maintained by the proprietor or contractor of an enterprise in order to put into practice the promotion of safety. The plant management or contractor should, as part of a safety promotion program, develop a procedure to recognize and acknowledge good safety performance by individuals, teams, sections, departments, or the organization.

Safety Promotion Approaches

Promotion of Safety in Meetings

A meeting can provide a good opportunity for promoting safety. Meetings suitable for promoting safety include orientation meetings for newcomers, training meetings, problem-solving meetings such as quality management meetings, information meetings such as toolbox meetings, and regular safety meetings. Safety and health films/videos of relevant subjects should be selected and screened for the benefit of those in the meetings and time should be allowed for the discussion of the subject after the viewing.

Promotion of Safety in Print

1. *Safety handbooks and brochures*: Up-to-date safety handbooks and brochures should be issued to increase safety awareness and as part of the safety training for staff.
2. *Safety bulletins and newsletters*: Safety bulletins and newsletters should be published by the enterprise to promote safety. They should contain interesting articles including how critical jobs executed in shutdown, pictorial presentations of safety rules, safety procedures and safe systems of work, a staff suggestion scheme, to encourage readership.
3. *Safety notice boards*: The boards should be used to post safety policy, rules, news, suggestions, accident reports, accident reporting procedures, emergency procedures, circulars, memos, notice of safety video shows and drills, etc.
4. *Safety billboards*: Vital statistics (such as up-to-date safety performance statistics of the enterprise and the individual work site including up-to-date accident-free days, no-loss days, or number of accidents) should be posted at the main entrance to the workplace or a conspicuous position within the workplace (such as the site office entrance).
5. *Safety posters*: Safety posters should be posted at strategic locations in the workplace (such as the main entrance and the production line). They should be weatherproof, relevant, up-to-date, clear, and attractive.
6. *Promotion of safety with awards and recognition*: The purpose of safety awards is to recognize and promote safe work practices and reinforce positive attitudes toward safety. An award does not have to be large but it should be meaningful and appropriate for the occasion and the employee or a group of employees.

11

Communication Package

Communication is the key link between people, ideas, and information. Shutdown communications management centers on determining who needs what information and when—and then produces the plan to provide the needed information. The communications management includes generating, collecting, disseminating, and storing communication. Successful shutdown execution requires effective and efficient communication.

Shutdown communications management includes three processes, which may overlap with each other. The three processes are

1. *Communication planning*: The shutdown manager will need to identify the stakeholders and their communication needs and determine how to fulfill their requirements.
2. *Information distribution*: The shutdown manager will need to get the correct information on the correct schedule to the appropriate stakeholders.
3. *Performance reporting*: The shutdown manager will rely on other performance measurements to create status reports, measure performance, and forecast shutdown conditions.

A communications management plan can provide:

- A system to gather, organize, store, and disseminate appropriate information to the appropriate people: The system includes procedures for correcting and updating incorrect information that may have been distributed.
- Details on how needed information flows through the shutdown to the correct individuals: The communication structure documents where the information will originate, to whom the information will be sent, and in what modality the information is acceptable.
- Specifics on how the information to be distributed should be organized, the level of expected detail for the types of communication, and the terminology expected within the communications.

- Schedules of when the various types of communication should occur. Some communication, such as status meetings, should happen on a regular schedule; other communications may be prompted by conditions within the shutdown.
- Methods to retrieve information as needed.
- Instructions on how the communications management plan can be updated as the shutdown progresses.

Performance reporting covers more than just cost and schedule, though these are the most common concerns. Another huge concern is the influence of risks on the shutdown success. The shutdown manager and his or her team must continue to monitor and evaluate risks, including pending risks, and their impact on the shutdown success.

Performance reporting involves six things:

1. *Status reports*: How's the shutdown right now?
2. *Progress reports*: How complete is the shutdown? How much more work remains?
3. *Forecasting*: Will this shutdown end on schedule? Will the shutdown be on budget? How much longer will this shutdown take? And how much more money will this shutdown need to get finished?
4. *Scope*: How is the shutdown scope? Is it doable within the time frame?
5. *Quality*: What are the results of quality audit, testing, and analysis?
6. *Risks*: What risks have come into the shutdown and what has been their effect on the shutdown?

There are many approaches you can take in communicating and many media to choose from.

Methods of Communicating

Some contend that e-mail is an overused medium for communication. It's certainly convenient—at least for most people. And it ordinarily reaches widely dispersed and distant recipients quickly—although there's no guarantee that the recipient will read the message right away.

Telephone: Calling someone on the phone can be an immediate (if they answer!), interactive method of communicating without creating a permanent, written record. Phone conversations allow you to hear voice inflections, although obviously you cannot view body language or other nonverbal communication.

Handwritten notes: Probably the most informal of all communication methods, short notes written by hand are an excellent way to provide positive recognition. Although they usually take very little effort, they convey "the personal touch" much more than verbal approaches or formal memos or e-mail messages.

Printed and mailed memos and letters: With the advent of email, memos and letters are now generally reserved for more formal or official communication. They're slow and one-sided, but good when formal signatures are required and a permanent record is desired. Hence, printed, mailed memos and letters are still used frequently in contractual situations.

Formal presentations: Formal presentations are often used in situations where the distribution of information may be enhanced by an explanation or the information is too complex for written documentation. Formal presentations are often done in a group setting, thus ensuring that everyone gets the same level of understanding. They allow for impressive graphical displays of information, but often require a lot of preparation. They're effective when you're trying to promote understanding, enlist support, or expedite a decision (e.g., management approval to proceed).

Conducting Meetings

Normally there will be two meetings daily. One is a meeting that the shutdown coordinator shall conduct with core team members including site in-charges of executing departments and operation departments for only half an hour. The other one is a detailed job review meeting held at different locations consisting of site engineers with operation shift in-charges to check progress and constraints in detail.

The main review meeting chaired by the shutdown coordinator is held daily at a specific location comprising core team members to discuss the progress versus the plan, the extra resource requirement, and the risk. Meetings can be a very effective way of getting a clear picture and resolving an issue across the table. They bring people together for a relatively short amount of time so that large amounts of information can be shared and problems resolved immediately.

The shutdown daily meeting is held for the following reasons:

- *Progress*
  - Review of actions from last meeting
  - Progress from last period
  - Forecast progress for next period

- *Decision*—to develop and agree upon a decision
- *Agreement*—to present a case on a decision and seek collective acceptance
- *Information*—to communicate information or decisions that have been made
- *Opinion*—to collect viewpoints and perspectives from participants
- *Instruction*—to provide direction, enhance knowledge, or teach a skill
- *Review*—to analyze some aspect of the shutdown, such as design or due to some constraints for which some jobs are postponed to the next opportunity
  - Problems/issues
  - Resource requirement

Turnaround progress meetings: The turnaround progress review meetings shall be at two levels: the steering committee and the working committee.

Steering committee meeting: These meetings shall be chaired by the head of the refinery. The meeting shall be held once in a week and minutes will be recorded and circulated to the Director of Refineries. The main objective of this meeting shall be

- Presentation by respective planner highlighting delays and giving expected date of completion of the turnaround
- Monitoring the turnaround progress
- Resolving any issues related to the turnaround

Working committee meeting: This meeting shall be chaired by the GMO. The meeting shall be held every day except for the day when the steering committee meeting is scheduled. The main objective of this meeting shall be

- Monitoring the turnaround progress
- Resolving any issues related to the turnaround

Communication for Success

Up-to-date information is the key to the success of a shutdown. Lack of job information during shutdown execution is the root of many shutdown failures. The status of all jobs must be communicated in a timely manner. A formal routine should be developed to report job progress.

1. Each morning, the shutdown coordinator should communicate with the shift coordinators. If a two-shift operation is being employed, the second (or night) shift coordinator should have already briefed the first shift coordinator.

2. During the day shift, all problems which may result in a delay in completion time or an increase in shutdown cost should be communicated.

3. At the end of every day shift, the shutdown planner should meet with all the shutdown support personnel and update the shutdown status.

4. Ask "What heavy equipment coordination is required today?"

It is important that individual daily schedules be broken out of the master schedule before each day of the shutdown. The functional team leader shall discuss this schedule with the site supervisor.

- Emphasis shall be given on writing in a logbook in detail along with start and completion time of each activity, in each individual section like mechanical, electrical, and civil which in turn will help the shutdown planner in the future. This will also facilitate the engineer in the next shift regarding the status of the job and in analyzing any delay/advancement. The shutdown schedule shall be regularly updated by entering the actual start/completion time.

- The detailed meetings regarding job progress and constraints that are held with shift coordinators, site engineers, core team members and operation department shift in-charges provide the background information. Each team leader should verify the statements made during these sessions and take necessary action to get the shutdown on track.

- The evening shift schedule can also be developed ahead of time. Any changes required to the evening schedule prior to the shift change should be minor. The maintenance first-shift supervisor can communicate these changes.

A walkie-talkie is required for easier communication during execution.

The communications management plan describes how stakeholders can access shutdown information between published data as well. You might consider setting up an intranet site for your shutdown and posting the appropriate shutdown documentation there for the stakeholders to access anytime they wish.

12

Shutdown Control

The duration of turnaround event is short and utilizes large manpower resources that include plant and maintenance personnel, technicians, craftsmen, and skilled and specialist contractors. The magnitude of the inter-related activities of the turnaround event requires stringent control and coordination.

Control can be defined as the process where the following actions can be done:

- *Tracking progress*: Knowing where you are, compared with where you're supposed to be. The shutdown manager will want to use a periodic reporting system that identifies the status of every activity scheduled for work since the last progress report. These reports summarize the progress for the current period as well as the cumulative progress for the entire shutdown.

- *Detecting variance from plan*: Variance reports are of particular importance to management. They are simple and intuitive, and they give managers an excellent tool by which to quickly assess the health of a shutdown. To detect variance, the shutdown manager needs to compare planned performance to actual performance. Exception reports, variance reports, and graphical reports give management the information necessary for decision making in a concise format.

- *Forecasting and evaluating potential problem areas*: This can be done prior to their occurrence so that preventive action can be taken.

- *Taking corrective action*: To take corrective action, it is necessary to know where the problem is and to have that information in time to do something about it. Once there is a significant variance from the plan, the next step is to determine whether corrective action is needed and then act appropriately. In complex shutdowns, this requires examining a number of what-ifs. When problems occur in the shutdown, delays result and the shutdown falls behind schedule. For the shutdown to get back on schedule, resources might have to be reallocated. In larger shutdowns, the computer can assist in examining a number of resource reallocation alternatives and help to pick the best.

- *Knowing what lies ahead that can affect you*: This action reviews trends or situations so that their impact can be analyzed and, if possible, an action proposed to alleviate the situation.
- *Providing constant surveillance of shutdown conditions*: This is useful so that a "no-surprise" situation is created effectively and economically.

The controlling process group is involved in taking measurements and performing inspections to find out if there are variances in the plan. If variances are discovered, corrective action is taken to get the shutdown back on track, and the affected shutdown planning processes are repeated to make adjustments to the plan as a result of the variances. Shutdown managers often make the same mistake when trying to keep control of their job. They get wrapped up in the here and now—the measurement and evaluation of their future situation is ignored.

What Are You Actually Controlling?

Two of the targets pertain to the consumption of resources:

1. *Schedule*: Was the shutdown completed *on time*?
2. *Cost*: Did the shutdown come in *at cost*?

The other two targets are tied to the deliverables of the shutdown:

1. *Functionality*: Do shutdown deliverables have the expected *capability*?
2. *Quality*: Do the deliverables perform as promised?

Detailed plans are created to satisfy two basic objectives: first, to provide a map for the shutdown team to follow during execution; and second, to provide you with an instrument you can use to evaluate whether or not the execution is staying on course as per plan. Your ability to evaluate progress, calculate your variance from the plan, and predict the future depends upon a number of key process elements. Among these are the following elements:

- A baseline of measurement
- Processes and methods for gathering data
- An ability to get good data
- An emphasis on timeliness
- Processes, tools, and methods for analyzing past, present, and future performance

What Are the Baselines?

The shutdown work has three baselines that will be used for performance and progress measurement. They are the scope baseline, the time or schedule baseline, and the cost baseline.

The *scope baseline* is the sum of the deliverables of the shutdown. It represents all the work that must be done to complete the shutdown. Any deliverables that are not included in the scope baseline will not be delivered to any of the stakeholders. Generally, however, it is better to establish the scope of the shutdown first and then, after the stakeholders have agreed on the scope, establish the cost and schedule. The scope baseline is the baseline from which all changes must be made. The current scope baseline begins with the original scope baseline; all changes that are approved are added or subtracted from the original scope baseline. After the changes are made, a new baseline is created, which becomes the current baseline. The performance measurement system and the progress measurement system will measure performance and progress against the current baseline. The cost and schedule baselines will probably need to be adjusted each time the scope baseline is changed.

The time baseline is the schedule of all the work that will be done to produce the scope baseline. Each item of work in the schedule is an item of work that is required to produce an output that either contributes to the delivery of a deliverable or that is an input required by another scheduled task in the shutdown. Scheduled tasks that do not contribute to the delivery of a deliverable or an input to another task should not be part of the shutdown schedule. So exactly what kind of information should you be gathering to evaluate your current variance and to maintain control of your shutdown regarding schedule?

- Date that each completed activity was scheduled to start and finish
- Date that each completed activity actually started and finished

The *cost baseline* is the budget of the shutdown. The cost baseline is derived from the estimates of the work to be done. It is usually shown as a cumulative estimated cost or budget curve of all the tasks and associated support—including people—that are expected to be needed during the life cycle of a shutdown. The cost estimates are determined from the work breakdown structure (WBS) and in conjunction with the schedule development. Resources and schedules have to be played against each other to obtain the most efficient and cost-effective baseline possible. The cost baseline does not include the contingency budget or the management reserve. The final or detailed cost estimates provide a baseline for assessing the financial performance of the shutdown. So exactly what kind of information should you be gathering to evaluate your current variance and to maintain control of your shutdown regarding cost?

- Estimated expenditure for all activities
- Actual expenditure (or labor hours logged) for each completed activity

Ideally, a shutdown should never vary from its original plan; therefore, a comparison between actual performance and the baseline should show no variance. In reality, this zero variance never happens.

Often, after planning is over and the shutdown begins, we find that

- Vital tasks and deliverables have been left out of the plan.
- The way we will do the work has changed. There is new discovery after opening the equipment.
- Estimates for cost, schedule, and resources are shown to be wrong.
- One or more stakeholders have asked for increases or decreases to the scope.

All of these are valid reasons to change the baseline early. In fact, a good way to handle this is to allow such changes for the first couple of reporting periods after the shutdown begins, then freeze the baseline for the remaining period.

Measuring Progress

The key to finishing a shutdown on time and on budget is to start out that way and stay on track throughout the shutdown. When shutdowns start with challenging schedules, if they fall behind, even by a little, they spend the rest of the shutdown trying to catch up. Other shutdowns, however, seem to have a self-correcting process built into them; if they fall behind a little, the problem is quickly identified and dealt with immediately. Progress measurements are the tools we use to identify problems when they are small and there is still time to catch up. Because cost and schedule progress comprise two-thirds of the cost–schedule–quality equilibrium, they are the primary focus of progress measurement.

Controls are actions taken as a result of reports. When implemented, controls are designed to bring actual shutdown status back into conformance with the shutdown plan. These reports are designed to support control activities by drawing attention to certain aspects or characteristics of the shutdown, such as planned versus actual schedule, trends in the schedule, and actual versus planned resource use.

We typically track performance levels, costs, and time schedules. There are three reasons to use reports in your shutdown.

Progress Reporting System

Once shutdown work is under way, you want to make sure that it proceeds according to plan. To do this, you need to establish a reporting system that keeps you informed of the many variables that describe how the shutdown is proceeding as compared to the plan.

A reporting system has the following characteristics:

- Provides timely, complete, and accurate status information
- Doesn't add so much overhead time as to be counterproductive
- Warns of pending problems in time to take action
- Is easily understood by those who have a need to know

There are different types of shutdown status reports:

Current period reports: These reports cover only the most recently completed period. They report progress on those activities that were open or scheduled for work during the period. Reports might highlight activities completed and variance between scheduled and actual completion dates. If any activities did not progress according to plan, the report should include a discussion of the reasons for the variance and the appropriate corrective measures that will be implemented to correct the schedule slippage.

Cumulative reports: These reports contain the history of the shutdown from the beginning to the end of the current report period. They are more informative than the current period reports because they show trends in the shutdown progress. For example, a schedule variance might be tracked over several successive periods to show improvement. Reports can be at the activity or shutdown level.

Exception reports: Exception reports record variances from the plan. These reports are typically designed for senior management to read and interpret quickly. Reports that are produced for senior management require special consideration. Senior managers do not have a lot of time to read reports that tell them that everything is on schedule and there are no problems serious enough to warrant their attention. In such cases, a one-page, high-level summary report that says everything is okay is usually sufficient. It might also be appropriate to include a more detailed report as an attachment for those who might wish to read more in detail. The same might be true of exception reports. That is, the one-page exception report tells senior managers about variances from the plan that will be of interest to them, while an attached report provides more details for the interested reader.

There are some reasons why you would want to measure duration and cost variances every week:

- *Allow early corrective action*: As suggested, the shutdown manager would prefer to be alerted to a schedule or cost problem early in the development of the problem rather than later. Early problem detection may offer more opportunities for corrective action than later detection.
- *Determine weekly schedule variance*: In our experience, we found that progress on activities open for work should be reported on a weekly basis. This is a good compromise on report frequency and gives the shutdown manager the best opportunity for corrective action plans before the situation escalates to a point where it will be difficult to recover any schedule slippages.

Analyzing the Schedule

Schedule analysis relies heavily upon graphic techniques. A basic shutdown control schedule shows three characteristics that are displayed for each activity:

The original baseline plan

The amount of progress made

The forecasted time-to-complete

A *Gantt chart* is one of the most convenient, most used, and easy-to-grasp depictions of shutdown activities that we have encountered in our practice. The chart is formatted as a two-dimensional representation of the shutdown schedule with activities shown in the rows and time shown across the horizontal axis. It can be used during planning, for resource scheduling, and for status reporting. The only downside to using Gantt charts is that they do not contain dependency relationships. Some shutdown management software tools have an option to display these dependencies, but the result is a graphical report that is so cluttered with lines representing the dependencies that the report is difficult to read. In some cases, dependencies can be guessed from the Gantt chart, but in most cases, they are lost.

Control Process

The shutdown control process is designed to spot problems early, while they are still small enough to correct. It is an iterative feedback loop in which the shutdown manager uses measurement and testing to evaluate deviations from the plan as to cost, schedule, quality, and risk. These deviations may or may not result in corrective action. The key is to monitor closely enough and often enough to spot such deviations before they get out of control.

There are five steps in the shutdown control process:

1. Define what will be measured and how often. This should incorporate business requirements, cost constraints, technical specifications, and deadlines, along with a preliminary schedule for monitoring that includes who is responsible for it.

2. Monitor progress and evaluate deviations from the plan. During each reporting period, two kinds of information are collected:

    a. Actual shutdown data, which include time, budget, and resources used, along with completion status of current tasks.

    b. Unanticipated changes, which include changes to budget, schedule, or scope, that are not results of shutdown performance. For example, heavy rain may delay the completion of a shutdown.

3. Report progress. Keep reports succinct and timely. Do not delay a report until after a problem is "fixed" to make the report look better. Likewise, avoid lengthy reports that delay the dissemination of important information to others in the organization.

4. Analyze the report. Look for trends in the data. Avoid trying to "fix" every deviation. If there is no trend to the deviation, it likely does not require corrective action at this time.

5. Take action where necessary. This includes updating the shutdown plan and notifying any stakeholders who are affected by the changes. If the changes are big enough, they will require steering committee approval in advance.

Where should the boundary line be drawn? There should be a predetermined tolerance limit—only variances outside that limit need be approved by the sponsor. If the tolerance is too small the shutdown manager would spend all his time getting tiny variances approved. If the tolerance is too wide, the sponsor has little or no control over costs. Before a stage starts the shutdown manager agrees on how much he can change things without seeking sponsor approval—i.e., his tolerance limits. The shutdown manager must then decide how much of his tolerance he will delegate to those below him in the chain of command.

Bear in mind also that tolerance is not on the current position versus the plan but on where we are going to end up. So, if we're currently 25% over budget but nevertheless confident of completing the stage on budget, no approval is needed. If we are currently on budget but think we will end up 25% over budget, that requires approval (unless we have a very generous tolerance limit indeed!).

Change Management

Change management is the process by which changes to the shutdown scope, deliverables, timescales, or resources are formally defined, evaluated, and approved prior to implementation. A core aspect of the shutdown manager's role is to manage change within the scheduled time successfully. This is achieved by understanding the business and system drivers requiring the change, documenting the benefits and costs of adopting the change, and formulating a structured plan for implementing the change. Issues and risks that crop up come under change management.

At the start of a shutdown, there is always uncertainty. However well the requirements are defined, they will change as the shutdown progresses. Issues will arise that were not foreseen.

What is an "issue" in the context of a shutdown? A risk that has happened, something unexpected that has occurred, a potential problem that has come to light—in other words, a threat to shutdown success has been identified. We need to deal with these threats.

What is the difference between an issue and a risk? Threats to shutdown success that are identified at the start of the stage or shutdown are called risks. Frankly it doesn't matter too much whether these things that crop up during the shutdown are called issues or risks. What does matter is that how they are managed.

- Before the shutdown starts, design an issue process that will meet your shutdown's need.
- When an issue arises, write down what it is.
- Assign the issue to somebody to resolve. In a small shutdown, the shutdown manager will decide who to give the issue to for resolution. In a large shutdown, this might be delegated to the team leaders—they decide which team member should be asked to resolve an issue.
- The person asked to resolve the issue does whatever it takes to get it resolved.
- Escalation paths must be open. The last thing you want to happen when the first major issue arises is not to know how to get to the sponsor—so you send him an email which sits in his inbox for a week. Before the shutdown starts, agree with the sponsor and steering committee members how you'll get urgent issues to them—via

their secretary, via a phone call? Get them to agree that they will make timely decisions.

- Write down how the issue was resolved.
- In status meetings, monitor open issues and give particular attention to issues that are past their target resolution date.

Let's have a look at a sample issue form.

Issue Form	
Job Name:	Issue No:
Short Description of Issue:	
Raised By:	Date Raised:
Description of Issue:	
Resolution Comments:	
Closed By:	Date Closed:
Issue Closing Agreed By:	Date:
Discussed in shutdown review meeting on date:	

Change requests (CR): What happens if continuous, unbridled change is allowed during a shutdown? The shutdown will go on for a very long time and may never finish. At the other extreme, allowing no change at all is a nice idea but would probably mean that whatever was delivered wouldn't meet the business need.

Never muddle the change budget with contingency. There should be a clear, separate budget for change with a clear owner. Contingency is therefore anything else that might crop up, though one reason for spending contingency could be that there is more change than expected and contingency is raided to fund it.

A CR process typically has three steps. In the first, the person raising the CR says two things: what change they would like and why—what is the business justification for disrupting this shutdown half way through; who raised the CR; and when was it raised.

Very simply someone can now make an informed, business-like decision: does the benefit of this change justify the date hit and the extra cost.

13

Quality Plan

Quality is one of the triple constraints found in all shutdown execution.

- It's the third leg to the successful completion of a shutdown and more typically defines whether stakeholder expectations were met.
- Being on time and on budget is one thing—if you deliver the wrong product or an inferior product, being on time and on budget suddenly don't mean much. If during start-up there is lot of trouble in handing over the plant and if you need to open the equipment for bad work, it has a great impact on the shutdown time. This could be one of them if you don't plan and monitor quality properly on your shutdown execution.

Quality planning is a process that is concerned with targeting quality standards that are relevant to the shutdown work and devising a plan to meet those standards. The quality management plan is an output of this process. It describes how the quality policy will be implemented by the management team during the course of the shutdown execution.

Producing a quality plan is not complex. It involves identifying all the deliverables at the start of the shutdown and deciding how to best validate their quality. There is an overhead in undertaking quality checks but this is offset by not having to fix things further down the line. Inevitably, the later you find a problem, the longer it takes to fix.

You need to decide how much focus you put on the quality of the shutdown, and how much on the quality of the deliverables.

- The shutdown quality refers to quality events, things like applying proper shutdown management practices to cost, time, resources, communication, etc. It covers managing changes within the shutdown.
- The deliverable quality refers to the "fit for purpose" aspect. It covers things like how well you did your inspection of equipments and how well you have attended to and fixed the problem.

Shutdown Quality

From a business perspective, shutdown quality is usually judged on the following criteria:

- Was the shutdown completed on time?
- Was the shutdown completed within budget?
- Did the system meet my needs when it was delivered?
- Does the system comply with corporate standards for such things as user interface, documentation, quality standards, etc.?
- Is the system well engineered so that it is robust and maintainable?
- Is it completed safely?

Deliverable Quality

Quality control in a shutdown typically involves insuring compliance with minimum standards of material and workmanship in order to insure the performance of the facility according to the design. These minimum standards are contained in the specifications described in the contract. With the attention to conformance as the measure of quality during the construction process, the specification of quality requirements in the design and contract documentation becomes extremely important. Quality requirements should be clear and verifiable, so that all parties in the shutdown can understand the requirements for conformance. Much of the discussion in this chapter relates to the development and the implications of different quality requirements for construction as well as the issues associated with insuring conformance.

Quality plan	A plan as to how and when "quality events" and "quality materials" are applied to a shutdown
Quality control	The implementation of the "quality events" in the "quality plan"
Quality assurance	QA is an umbrella term. It refers to the processes used within an organization to verify that deliverables are of acceptable quality and that they meet the completeness and correctness criteria established. QA does not refer to specific deliverables • The preparation of a "quality plan" for a shutdown is part of QA. • The development of standards is part of QA. • The holding of a "quality event" is part of QA.

Inspectors and quality assurance personnel will be involved in a shutdown to represent a variety of different organizations. Each of the parties directly concerned with the shutdown may have their own quality and safety inspectors, including the owner, the engineer/architect, and the various constructor firms. These inspectors may be contractors from specialized quality assurance organizations. In addition to on-site inspections, samples of materials will commonly be tested by specialized laboratories to insure compliance. Inspectors to insure compliance with regulatory requirements will also be involved.

Planning Quality

A quality plan needs to cover a number of elements:

- What needs to go through a quality check?
- What is the most appropriate way to check the quality?
- When should it be carried out?
- Who should be involved?
- What "quality materials" should be used?

What Needs to Be Checked?

Typically what needs to be checked are the deliverables. They are sometimes called metrics, and their purpose is to specifically describe what is being measured and how it will be measured according to the quality control plan and process.

Developing Checklists

Checklists provide a means to determine if the required steps in a process have been followed. As each step is completed, it's checked off the list. Checklists can be activity-specific. Sometimes, organizations may have standard checklists they use for shutdowns. You might also be able to obtain checklists from professional associations. Remember that checklists are an output of this process but are a tool and technique of the risk identification process, and are an input to quality control.

Quality Assurance during Shutdown in a Process Plant

- Quality assurance for flushing activities and system for condensate draining
- Blinding/de-blinding

- Water wash procedure
- Passivation procedure
- Chemical cleaning of equipment wherever possible prior to shutdown
- Online cleaning of furnace tubes
- Quality assurance system for hot jobs
- Quality assurance system for flange assembly
- Quality assurance system for repair/replacement
- Quality assurance system for material check
- Quality assurance system for nondestructive testing (NDT), for example, radiography, die penetrant testing(DPT), ultrasonic flaw detection, hydrotest, positive material identification (PMI), etc.
- Quality assurance system for stress relieving (SR)
- Compliance of statutory inspection requirements, for example, factory inspector, electricity authority, boiler regulation
- Review and updation of scope of inspection for material procurement
- Adherence to quality assurance system as per scope of inspection
- System of second check for completed jobs (logging, completions, and signature system)
- Preservation of idle process plants/equipment as per standard procedure
- Equipment handling procedure, for example, tube bundle, safety valve, control valve, gasket, etc.
- Identification, collection, segregation, and tagging of all types of gaskets

The only ultimate aim a quality program can have is to end up with perfection. Deciding what should be checked by whom is important.

Quality Assurance Plan
(Typical) Welding

| Sl. No. | Activity | Responsibility | | | Approved on | Remarks |
		Agency (%)	ML (%)	IP (%)		
1.	Material confirming to specification and review of test certificate	100	20	10		
2.	Selection of electrode from owner-approved manufacturer's list	100	10	10		
3.	Welding procedure document approval	100	50	100		
4.	Welder qualification	100	50	100		

Quality Assurance Plan
(Typical) Welding
(continued)

Sl. No.	Activity	Responsibility			Approved on	Remarks
		Agency (%)	ML (%)	IP (%)		
5.	D.P. check for root face (to check lamination, if any, for higher thickness)	100	10	100		
6.	Joint fit-up clearance (mark joint number and welder number)	100	10	100		
7.	Check electrode baking oven	100	90	10		
8.	Check electrode with respect to owner-approved manufacture list, batch number, expiry date, etc.	100	90	10		
9.	Checking baking cycle of electrodes		100	10		
10.	Use portable pot for electrode (to avoid moisture ingress)	100	100	10		
11.	D.P. test for root run (after gouging/grinding)	100	10	100		
12.	Maintain preheat temperature before starting welding	100	10	100		
13.	Maintain interpass temperature	100	10	100		
14.	Visual check of completed joint	100	10	10		
15.	Radiography of weld joints as per the percentage indicated in drawing	100	10	100		
16.	Check PWHT temperature recorder with respect to calibration certificate	100	100	100		
17.	Check heat treatment chart (initial by inspector at start and end of cycle on the chart)	100	10	100		
18.	Check PWHT hardness of weld joint	100	10	100		
19.	Final radiography of weld joint	100	10	100		

Quality Assurance Plan
(Typical) Column

Sl. No.	Checks	Executing Agency	Inspection by			Remarks
			PN	PS	IP	
1.	Cleaning of					
	Shell internal	ML	X		X	
	Tray internals	ML	X		X	
	Process nozzles	ML	X			

(continued)

Quality Assurance Plan
(Typical) Column
(continued)

Sl. No.	Checks	Executing Agency	Inspection by			Remarks
			PN	PS	IP	
	Instt. connections	ML/IT	X			
	Orifice assembly. In over flash line	ML/IT	X			
	Coke trap	ML	X			
	Demister	ML	X	X		
2.	Repair/replacement (including hot jobs)					
	Shell	ML	X		X	
	Shell lining				X	
	Tray support ring				X	
	Nozzle/RF Pad				X	
	Internal support structure				X	
	Tray segments				X	
	Down comer				X	
	Replace demister pad				X	
3.	Thickness survey					
	Shell				X	
	Tray internals				X	
4.	Thoroughness of corrosion inhibiter/chemical dosing	ML	X	X		
5.	Tray assembly					
	Alignment of tray	ML	X	X		
	Valves, bubble cap, etc.	ML	X	X		
	Clamp, washes, bolts	ML	X	X		
	Downcomer	ML	X	X		
	Seal-pan leak test	ML	X	X		
6.	Distribution header assembly					
	Flange/gasket joints	ML	X			
	Threaded joints	ML	X			
	Fixing of nozzles	ML	X			
7.	Packing pad					
	Condition of packing		X	X		
	Height of packing		X	X		
	Bed limiter		X	X		
8.	Chimney trays		X	X		
9.	Thermowell connections	ML/IT	X	X	X	
10.	Fixing of tray manways	ML	X	X		
11.	Manhole box-up	ML	X	X		
12.	Insulation	CL	X		X	
13.	Blind/de-blind of nozzles with correct gasket and fasteners	ML	X		X	
14.	Painting	LC			X	

Quality Assurance Plan
(Typical) Column
(continued)

Sl. No.	Checks	Executing Agency	Inspection by			Remarks
			PN	PS	IP	
15.	Insulation	CL			X	
16.	Fire proofing	CL			X	

Note: While boxing up flange joints, the length of studs/bolts should be such that a
minimum of two threads are shut down out of the nut on either sides.
Agency - who will do the job
ML - mechanical engineer
IP - inspection engineer

Quality Assurance Plan
(Typical) Exchanger

Sl. No.	Checks	Executing Agency	Inspection by			Remarks
			PN	PS	IP	
1.	Cleaning of					
	• Shell, channel box, covers	ML	X			
	• Tube bundle	ML	X			
	• Flange faces	ML	X			
	• Shell/tube side nozzles	ML	X			
2.	Thickness survey					
	• Shell, channel box, plate, etc.	ML			X	
	• Shell/tube side nozzles	ML			X	
3.	Repair (including hot jobs)					
	• Weld deposit/patch repair on shell	ML			X	
	• Partition plate	ML			X	
	• Baffles, impingement plate	ML				
4.	Sand blasting/painting of waterside components, e.g., channel box, floating cover, and channel cover	CL			X	
5.	Replace sacrificial anode, wherever required	ML			X	
6.	Hydrotest					
	• Shell side	ML			X	
	• Tube side	ML			X	
7.	Use of correct gasket with respect to shape, size, and metallurgy	ML			X	
8.	Blind/de-blind with correct gasket and fasteners	ML			X	
9.	Painting	CL			X	
10.	Insulation (hot/cold)	CL			X	

Note: While boxing up flange joints, the length of studs/bolts should be such that a minimum
of two threads are shut down out of the nut on either sides.

Quality Assurance Plan
(Typical) Vessel

Sl. No.	Checks	Executing Agency	Inspection by			Remarks
			PN	PS	IP	
1.	Cleaning of					
	• Shell internal surface	ML	X		X	
	• Nozzles/tappings	ML/IT	X	X	X	
	• Internals, e.g., demister, raching ring	ML	X	X		
	• Distributor assembly	ML	X			
2.	Electrical items					
	• Transformer	EL	X	X		
3.	Thickness survey					
	• Shell	ML			X	
	• Nozzles	ML			X	
	• Steam coil	ML			X	
	• Distributor pipes	ML			X	
4.	Fixing installation of internals					
	• Demister	ML	X	X		
	• Raching ring, activated charcoal, etc.	ML	X	X		
	• Distributor pipe assembly. With nozzles	ML	X			
	• Steam coil	ML	X			
5.	Repair, hot job, etc.	ML			X	
6.	Manhole box-up	ML	X			
7.	Blind/de-blind with correct gasket and fasteners	ML	X		X	
8.	Painting	CL			X	
9.	Insulation (hot/cold)	CL			X	
10.	Fireproofing	CL			X	

Note: While boxing up flange joints, the length of studs/bolts should be such that a minimum of two threads are shut down out of the nut on either sides.

How might we get all shutdowns to include appropriate quality-checking activities in their plan? One way would be to have a rule that at the start of each stage the shutdown manager must review his plan with shutdown support, and shutdown support must agree that appropriate quality tasks have been built into the plan. And a further rule might be that the sponsor will not give the go ahead unless shutdown quality is satisfied.

14

Risk in Shutdown

A shutdown includes some element of risk. It's natural to be optimistic that you will overcome any unplanned event during the execution of a shutdown. The old adage, "optimism blinds—pessimism paralyzes" applies very well to shutdowns. The shutdown risk management strives to achieve a balance between optimism and pessimism by confronting potential risks during the shutdown planning phase.

Risk and uncertainty are unavoidable in shutdown life and it's dangerous to ignore or deny their impact. Adopting a "can do" attitude may be a good way to get your team members energized and committed, but it's a foolhardy approach when it comes to managing a complex shutdown.

Why Risk Analysis Is Necessary?

Risk analysis is conducted so that a shutdown team can accurately predict the shutdown completion date. It is based on a task-by-task analysis. We discovered that we can identify each task's expected duration that will be good most of the time—over 50% of the time if the task were repeated over and over. And, we can identify the likely worst case overrun for each task. We discovered a way to use this information to average out task overruns to create a shutdown time buffer. We create a time buffer for the Gantt chart to protect our completion date estimate from the tasks that overrun most of their time forecast.

The uncertainty in undertaking a shutdown comes from many sources and often involves many participants in the shutdown. As each participant tries to minimize its own risk, the conflicts among various participants can be detrimental to the shutdown.

What Is Risk Management?

A risk is a possible unplanned event. Because risks are the unpredictable part of the shutdown, it is important for us to be able to control them as much as possible and make them as predictable as possible.

Risk management is the process of identifying, analyzing, and quantifying risks, responding to them with a risk strategy, and then controlling them.

What Are the Basic Steps in Risk Management?

There are usually four steps considered in managing any risk.

1. Risk identification
2. Risk quantification
3. Risk response
4. Risk control

What Is Risk Identification?

Risk identification is the process of identifying the threats and opportunities that could occur during the life of the shutdown along with their associated uncertainties. The first step in risk management is identifying the risks that we will see in our shutdown. These are the things that threaten to stop us from delivering what we have promised on the schedule we promised for the budget we promised.

The first thing we must do in risk identification is recognize the areas of the shutdown where the risks can occur. This means that we will have to investigate the following areas:

- *Scope*: We must look at the work of the shutdown. The work breakdown structure (WBS) will be useful here. The shutdown scope must be clearly defined in terms of both the deliverables and the work that must be done to deliver them. Errors and omissions on the part of the shutdown team and the stakeholders must be minimized. As always, the WBS will be very helpful in doing this.
- *Time*: Estimates for the duration of the shutdown and the duration of the shutdown tasks must be done accurately and reliably. The sequence of work must be identified, and the interrelationships between the tasks must be clearly defined.
- *Cost*: Estimates for tasks must be done accurately and reliably. All associated costs must be considered and reported accurately.

Life cycle costs should be considered as well as maintenance, warranty, inflation, and any other costs.

- *Customer expectations*: Estimates of shutdown success must be considered in terms of customer needs and desires. The ability of the shutdown to be scaled up or manufactured in different quantities or for different uses and sizes must also be considered.
- *Resources*: This involves the quantity, quality, and availability of the resources that will be needed for the shutdown. Skills must be defined in the roles that will be necessary for the shutdown.

The following items can be explored to stimulate discussion and flesh out hidden risks:

- *Staffing assumptions*: Some important activities may depend on the attendance of essential personnel. Identify people who will be indispensable during the shutdown execution because of their special skills or knowledge. Next, determine whether or not the shutdown can proceed without these people. Also consider the potential for work stoppages or slowdowns.
- *Estimate risks*: Identify time and cost estimates that were developed with minimal information.
- *Procurement problems*: Any deliveries expected during the shutdown execution should be reviewed for potential delays or even cancellation. Items from sole-source suppliers represent the highest risk.
- *Shutdown files*: Review previous shutdown results. Even shutdowns performed in a different area can provide some insight into potential problems.
- *Commercial data*: Review trade articles for insight into some problems that others have encountered. The American Society of Professional Estimators has published articles on shutdowns and major modifications performed in industry.
- *Shutdown team knowledge*: Query the shutdown team. Team member recollections of previous shutdowns are useful, although less reliable than documented results.
- *Possible weather conditions*: Major storms, or just rain and snow, can affect the schedule of a shutdown.
- *Nature of the shutdown*: Sometimes the magnitude of internal damage is unknown. Corrosion, abrasion, or wear may be higher than expected. A similar risk exists when primary activities of a shutdown are performed for the first time. Application of new technologies or methods falls into this category.

What Is Risk Quantification?

Risk quantification is the process of evaluating the risk as a potential threat or opportunity. We are mainly concerned about two items: risk probability and risk impact. Risk probability tells us the likelihood that the risk will take place, and risk impact is the measure of how much pain or happiness will result if it does take place. Risks that have very high impacts with very low probabilities and risks that have very low impacts with high probabilities are usually of little concern, so we need to consider the combination of these two items before considering how important a risk is. The combination of impact and probability is called severity.

The objective of quantification is to establish a way of arranging the risks in the order of importance. In most shutdowns, there will not be enough time or money to take action against every risk that is identified. Two questions should be asked of the shutdown team for each item on the list: Is the probability that this risk will be encountered high or low? Would an occurrence of the event significantly lengthen the shutdown?

What Are Risk Response Strategies?

A risk response strategy is really based on risk tolerance, which has been discussed. Risk tolerance in terms of severity is the point above which a risk is not acceptable and below which the risk is acceptable.

There are many reasons for selecting one risk strategy over another, and all of these factors must be considered. Cost and schedule are the most likely reasons for a given risk to have a high severity.

There are basically five categories of classic risk reduction strategies: accepting, avoiding, monitoring, transferring, and mitigating the risk. Let us look at these in detail.

1. *Accepting the risk*: Accepting the risk means you understand the risk, its consequences, and probability and you choose to do nothing about it. If the risk occurs, the shutdown team will react. This is a common strategy when the consequences or probability that a problem will occur are minimal. As long as the consequences are cheaper than the cure, this strategy makes sense.

2. *Avoiding the risk*: You can avoid a risk by choosing not to do part of the shutdown. Changing the scope of the shutdown might change the business case as well, because a scaled-down product could have smaller revenue or cost-saving opportunities. "Risk/return"

is a popular expression in finance—if you want a high return on an investment, you will probably have to take more risk. Avoiding risks on shutdowns can have the same effect—low risk, low return.

3. *Monitoring the risk and preparing contingency plans*: Monitor a risk by choosing some predictive indicator to watch as the shutdown nears the risk point. For example, if you are concerned about a subcontractor's performance, set frequent status update points early in the shutdown and inspect his or her progress. Contingency plans are alternative courses of action prepared before the risk event occurs. The most common contingency plan is to set aside extra money, a contingency fund, to draw on in the event of unforeseen cost overruns.

4. *Transferring the risk*: Even though paying for insurance may be expensive, assuming all the risks yourself could cost a great deal more. Many large shutdowns purchase insurance for a variety of risks, ranging from theft to fire. By doing this, they have effectively transferred risk to the insurance company in that, if a disaster should occur, the insurance company will pay for it.

5. *Mitigating the risk*: Mitigate is jargon for "work hard at reducing the risk." The risk strategy includes several ways to mitigate, or reduce, the productivity loss associated with using a new software tool. Mitigation covers nearly all the actions the shutdown team can take to overcome risks from the shutdown environment.

Develop Response Plans

Understanding a problem is the first step in solving it, so clearly understanding each risk is the first step in risk response development. Defining a risk consists of describing specifically what problem might occur and the potential negative impact of the problem. Notice that in defining risk, we use terms such as *might occur and potential impact* because we are forecasting the future. So another element of risk definition is quantifying the probability that the risk event will actually take place. The combination of the probability of the risk and the damage it will cause helps prioritize the risks and determine how much we are willing to spend to avoid or reduce the risk. Every risk response attempts to reduce the probability and/or impact. The primary outcome of developing response plans is a risk log. The risk log is the full list of risks the shutdown team will actively manage and contains the specific tasks associated with managing each risk.

Establish Reserves: Ideally, some reserves have already been allocated in the shutdown's budget for responding to risks. However, once detailed risk planning has been performed, we will have a much more accurate assessment of how much money to set aside for responding to known risks and unknown risks. Contingency reserves are allocated for known risks. Management reserve is allocated for unknown risks.

Continuous Risk Management: Continuous risk management is a conscious repetition of risk identification, response development, and carrying out risk plans. At regular intervals, the known risks are reassessed and the team vigorously searches for new risks. The risk plan is updated to record whether the risks actually occurred and whether the response strategy actually worked. Reports to management include updates on the status of high-profile risks and the amount of contingency and reserve expended to date.

Why Do We Have Two Separate Reserves to Take Care of the Risks?

There are two kinds of reserves set up to budget for risks: the contingency reserve and the management reserve. The contingency reserve contains the money to do the risks that were identified. The management reserve contains the money to do the risks that were not identified. These two reserves are separated in order to have more control over how they are spent. Use of the two reserves usually requires one additional level of management approval.

Risk Control

Executing the risk management plan will obviously begin when an identified risk event occurs. However, even the most thorough review of potential events cannot identify all risks, so a plan of action may have to be implemented.

Contingency planning: Contingency plans are specific actions that are to be taken when a potential problem occurs. Although they're intended to deal with problems only after they've occurred, *contingency plans should be developed in advance.*

This helps ensure a coordinated, effective, and timely response.

In addition, some plans may require backup resources that need to be arranged for in advance. Contingency planning should be done only for the high-threat problems that remain after you've taken preventive measures.

Never forget the relationship between cost, schedule, quality, and risk. The so-called *triple constraint* is well known in shutdown management: cost, schedule, and quality are related because it takes time and money to produce a product. Risk is an inherent, though often unrecognized, factor in that relationship. Most strategies to cut schedule and cost increase risk.

When an executive asks a shutdown team to produce a more aggressive schedule (i.e., cut some time off the team's proposed schedule) or to sharpen their pencil to reduce the budget estimate, that executive is asking the team to add risk.

Maintain both a contingency and a reserve within the shutdown budget.

Contingency accounts for known risks and the possible cost of dealing with them if they arise.

Using a Risk Profile

One of the best ways to ensure the success of a shutdown is to apply the lessons learned from past shutdowns. This is done by using a risk profile. A risk profile is a list of questions that address traditional areas of uncertainty on shutdowns.

Good risk profiles follow these basic guidelines:

- *They are industry-specific.* For example, building an information system is different from building a shopping mall.

- *They are organization-specific.* While industry-specific profiles are a good place to start, the profiles are even better when they address risks specific to a company or department.

- *They address both product and management risks.* Risks associated with using and developing new technology are product risks. Management risk addresses shutdown management issues, such as whether the team is geographically dispersed.

- *They predict the magnitude of each risk.* Even simple, subjective indicators of risk such as "high—medium—low" contribute to a clearer assessment of specific risk factors. More specific quantitative indicators offer the opportunity for greater refinement and accuracy over many shutdowns.

Risk profiles are generated and maintained by a person or group independent of individual shutdowns. The keeper of the risk profile participates in postshutdown reviews to learn how well the risk profile worked and to identify new risks that need to be added to the profile. These profiles, when kept up to date, become a powerful predictor of shutdown success. The combined experience of the firm's past shutdowns lives in their questions.

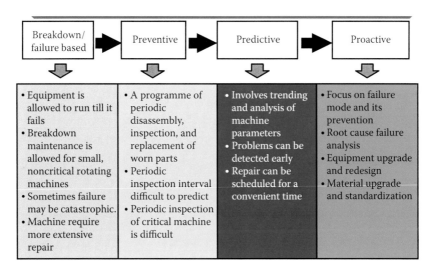

FIGURE 1.1

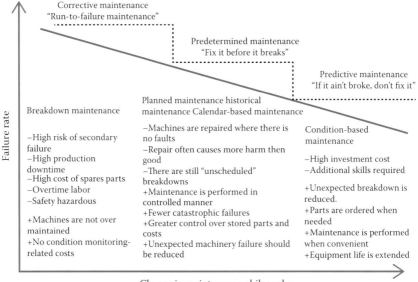

FIGURE 1.2

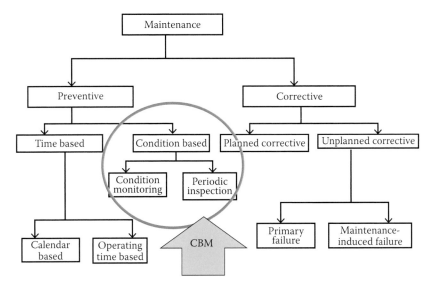

FIGURE 1.3

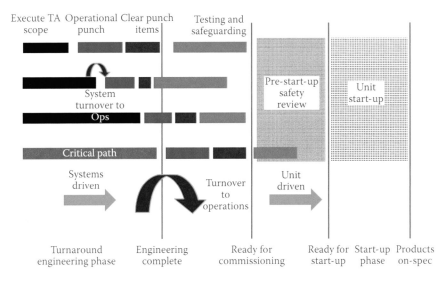

FIGURE 16.1

Historical Records

History continues to be the best predictor of the future. In addition to the history incorporated in the risk profile, a shutdown manager can investigate what happened on similar shutdowns in the past:

- Planned and actual performance records that indicate how accurate the cost and schedule estimates were.
- Problem logs that portray the unexpected challenges and relate how they were overcome.
- Postshutdown reviews that generate the lessons learned from the shutdown.
- Find out what a similar shutdown team did right and wrong, and learn from their experience.

15

Joint Integrity

Whether it is a pipe flange, valve bonnet, heat exchanger, or reactor manway, joint integrity relies not only on the mechanical design of the flange and its components but also on its condition, maintenance, and assembly. Piping system leakage is a common occurrence at process plants across many industries around the world and planned maintenance shutdowns are all too often followed by leaking joints on start-up.

Leaking joints are both costly—in terms of lost product and inefficient plant operation, downtime and repair costs, and potentially damaging or dangerous with safety and environmental consequences, not to mention the negative impact on corporate image. Where critical joints are concerned— that is, where leakage would cause plant shutdown, the process to be affected or danger to personnel or equipment—failure can be costly in many senses and integrity is particularly crucial. Further, achieving a leak-free start-up after a scheduled shutdown will avoid delays, reduce equipment and testing costs, avoid rework, and enable earlier demobilization of labor.

Plant personnel are looking for high joint integrity—leak-free joints with reduced numbers of incidents during start-up. To achieve joint integrity, a broader view of the bolted flange joint as a dynamic system must be adopted. A process is to be followed that manages the key elements of the bolted system, allows the design potential of the bolted joint to be realized, and helps in achieving continued leak-free operations.

A key component of these activities has been the development of guidelines for establishing an effective fluid-leak management program.

Managing Integrity

While there is a growing acceptance of the need to manage joint integrity as a key tenet of good maintenance practice, what is often not realized is the level of engineering and management required to ensure leak-free performance. Simply installing a gasket and tightening the bolts will not ensure a reliable leak-free joint.

A range of criteria will affect the level of management required for any one critical joint, from its physical size and operating pressure and temperature, to factors such as any fluctuations in temperature that it may be subjected to.

Causes of leaking flanged joints vary, but as a general rule, flange distortion, sealing surface damage, inappropriate gasket selection, incorrect bolt loads, and uncontrolled tightening methods are typically among the primary ones.

Work that needs to be undertaken during shutdown will typically include ensuring an appropriate surface finish, flatness and condition of the existing gasket face, including any re-machining as required. The rougher the surface finish, the higher are the bolt loads required to obtain a seal, for example, while any marks or defects greater than 30% of the flange sealing face width will be difficult to seal and should therefore be re-machined. Re-machining should also be considered if the flatness of the face is outside the maximum tolerance. Alternatively, if a new gasket is required, this will be inspected and installed. This is then followed by flange alignment (significant misalignment of the flange holes can require an additional load to overcome this) and controlled bolt tightening to the determined load using hydraulic tensioning or torquing. Bolt tensioning is generally accepted as the most accurate method of tightening—a technique that makes use of advanced hydraulic technology to induce accurate bolt stresses without creating torsional or bending stress. The bolt is gripped and stretched axially to the predetermined load using hydraulic pressure. Beneficially, because the stud is axially loaded no bending or torsional stress is induced, and as friction is an insignificant factor in the technique, repeatable and accurate residual bolt loads to specific requirements are obtained, and can be readily reproduced. The residual stud tension can be confirmed by ultrasonic or mechanical stress measuring equipment. A large number of tensioners can be used simultaneously to keep time to a minimum, and can be readily applied even in areas of difficult access, thanks to the design of modern strong yet compact and lightweight equipment which can meet the most stringent requirements and enable even the largest bolt sizes to be tightened to specific design requirements without resorting to wrenches or spanner extensions. On the other hand, where tensioning is not required or hydraulic tensioning equipment cannot be used, torque tightening (involving turning the nut to stretch the bolt) offers a simple and safe method of ensuring controlled tightening and loosening of bolts. A wide range of light, compact, safe, and user-friendly hydraulic torque tools and a complete range of wrenches are used for a torque load of 80,000 ft lb or 108 Nm.

Major Causes of Joint Leakage

The causes of leaks from pipe flange joints are

- Gasket problems: reused gasket; wrong size or pressure rating; wrong filler, metal strip, or compression stop ring materials; gasket damage during storage; absence of inner ring when needed to accommodate smooth flange surfaces; inadequate centering in four bolt flanges

- Flange surface condition problems: surface not 125–250 µin. rms; surface has gouges or scratches exceeding ASME B16.5 limits; flanges not flat within 0.015 in.

- Bolting problems: low strength bolts or soft nuts/washers; corroded threads; painted threads; stretched threads; failure to use hardened steel washer when flange surface is poor

- Misalignment and distortion problems: excessive lateral or angular misalignment

- Preload problems: too low a preload (gasket not compressed to a hard joint condition); too high a preload (yields bolts, yields flange, crushes gasket, excessive flange rotation); failure to lubricate threads and nut-to-flange interface; loss of preload during operation (gasket creep, thermal transients)

- Assembly problems: cocking due to lack of snugging joint before applying first pass; failure to follow cross-pattern loading; too few passes; too few leveling passes

Corrective Action for Guideline and Responsibility Deficiencies

These were human performance issues that could be corrected with proper training.

A decision was made to appoint a single point of responsibility for bolting methods and procedures. Thus came the appointment of a "bolting engineer," a collateral duty for a staff engineer. The assignment of this sole responsibility proved to be a key factor in leakage reduction. It resulted in the development of a bolting expert who became the central point of contact for bolting and leak-related questions.

In order to provide the necessary support to the facility, the assigned individual must be highly knowledgeable in the design, maintenance, and behavior of bolted gasketed joints and bolting in general.

Initially, it was anticipated that existing solutions to bolting problems were readily available and that the development of a bolting program would be a brief research and implementation process. The reality was that existing solutions were scattered, poor, or nonexistent.

Plant Staff Technical Knowledge Related to Leak Repair

Overall, plant personnel had limited technical training related to bolted joint behavior or gasketed joint performance. Individuals, departments, or groups were doing the best they could with what they knew.

Bolted joint assembly was based on bolt or stud size and standard torque tables with little or no regard to the type of joint, bolt material, flange design, gasket considerations, or required bolt preload or service conditions. In many situations, this approach is satisfactory. In many, if not most, others, it is totally inadequate.

Historical Records of Leaks and Maintenance

Records of leak repairs, if they existed, contained little useful information on the as-found conditions, cause of the leak, or specific corrective action to prevent recurrence. In the absence of historical data, the same ineffective repairs or maintenance activities were being performed on leaking joints.

Insufficient emphasis was placed on gathering data with which to perform root cause determination of the leaks. Root cause determination was rarely performed prior to planning and implementing repair actions. It was usually performed some time later to meet an arbitrarily assigned due date.

Historical Records of Corrective Action

This system has become a historical archive of problems and the corrective actions that were implemented. If a failure occurs, the component ID is entered into the system and a listing of prior problems is made available. Ineffective corrective actions quickly become evident.

Procurement of Bolting-Related Supplies

Procurement of gaskets was based solely on price with little input from maintenance or engineering personnel. The purchasing department freely substituted manufacturers and products if they were approved as "equal." "Equal" gaskets from various vendors and manufacturers were placed in the same stock bins under the same stock numbers. No attention was given to the fact that gaskets from different sources were not sufficiently identical and that they performed differently in the field or in specific joints.

No inspection criteria were employed for bolting or gasket materials except for checking that the paperwork was complete.

Joint Assembly Procedure

It is important to follow a controlled assembly sequence to ensure high joint integrity.

The following is a brief summary of the recommended procedure for joints assembled by torquing:

- Inspect all joints during disassembly for conditions that could lead to leakage.
- Obtain and inspect the new gasket; clean and inspect the flange on the gasket seating surface and where the nuts contact the flange; and clean and inspect the bolts, nuts, and washers.
- Lubricate the bolt on the threads, nut-to-flange interface, and hardened steel washers if used.
- Tension the joint.
- Tighten the nuts finger tight.
- Torque the nuts to about 5% of the final torque in a cross pattern to ensure that all clearance is out of the joint before starting the main torque sequence.
- Torque the bolts to the final specified torque in a cross pattern using three passes (33%, 67%, and 100%) for most joints and four passes (25%, 50%, 75%, and 100%) for critical joints, problem joints, misaligned joints, and where the bolt preload stress will be greater than 52,500 psi.
- For misaligned joints, inspect to ensure that the flanges are parallel within 0.015 in. after three leveling passes at 50% of the final specified torque.
- Finally, torque bolts sequentially until there is no further nut rotation. If more than three leveling passes are required, this is evidence that the gasket is still not fully compressed and consideration should be given to going to the higher 60,000 psi stress and torquing the joint again after 24 h.

Most industrial plants have a significant number of high-risk flanges that warrant plant-wide concern. Generally, these fall into one or more of the following categories:

- Flanges that CHRONICALLY leak—Certain flanges are a continuing problem because of their tendency to leak in spite of ongoing efforts to fix them.
- Flanges that MUST NOT leak—Certain flanges pose such potentially serious problems that they should never leak. Such leaks could cause:
  1. Physical injuries.
  2. Unscheduled work stoppages—production loss.
  3. Damage to plant or equipment.

4. Late schedules.

5. Fires.

6. Regulatory and/or legal problems.

7. Problems resulting from leaks in flanges can range from local in severity to plant-wide catastrophe. Although the range of negative results can vary widely, these high-risk leaks have one thing in common—ALL LEAKS ARE PREVENTABLE.

Leaks are symptoms of a loss of clamping pressure on the sealing element under operating conditions. This can be caused by a single factor or by a combination of factors.

Quality control factors that can negatively affect joint integrity include

- Variation or errors in bolt loadings during assembly procedures
- The flatness and/or texture of the flange face
- Bolt metallurgy
- The creep relaxation of the gasket material

Conclusion

Joint integrity is today firmly on the map, and is recognized as a crucial element of any plant maintenance program. There is no doubt that effective management of critical pressure-containing joints can reap rewards in terms of cost-savings and operational efficiency, not to mention removing risk. Moreover, with a leak-free start-up guaranteed to users of the management program, for example (an approach to eliminate leaking joints), the advantages of investing in such management programs become even greater.

16

Commissioning

Plant commissioning is the final phase of the shutdown and it is one of the most critical stages of a successful shutdown completion as it confirms and verifies whether the newly installed facility would be able to meet the performance criteria and established goals. On completion of the shutdown, the plant should return to a fully operational state with the absolute minimum of problems, ensuring trouble-free operation thereafter. The success of a shutdown can be measured by the problems experienced during start-up. However, this phase sometimes receives little attention with short upfront planning from the contractor due to other competing priorities until it is time to start the plant up. In some cases, the contractor's forces have left the site resulting in repeated trips.

Before a *schedule start-up* after a shutdown of the unit, one must ask a number of questions that must be answered in the affirmative. Some of them include

- Has all of the work been completed?
- Has the work been completed according to the original design? If not, then is change of design documentation available for review?
- Has the work been inspected?
- Have relevant tests been conducted on the process equipment?
- Has the documentation and instructions for new equipment been prepared?
- Have the operators and technicians been trained on any new equipment that has been added to the process?
- Has the instrumentation been checked?
- Are the process analyzers working correctly?
- Have control valves been stroked?
- Have the electrical systems been tested and inspected?
- Have all pressure tests and leak tests been completed?
- Has the piping been connected, inspected, and blanks removed?

- Have relief valve installations been inspected and any blanks removed?
- Has all scrap been removed from the area?
- Is area housekeeping in good order?
- Has start-up documentation been prepared for the operators?
- Have upstream and downstream process unit managers been notified of the start-up?
- Have the electrical, steam, and cooling water utility managers been notified of the pending start-up?
- Has a pre-start-up safety review (PSSR) been conducted?

The different phases in plant commissioning and start-up are

- Phase A: Prepare and plan equipment for precommissioning/mechanical testing
- Phase B: Clean and pressure test systems
- Phase C: Check and prepare major mechanical equipment, instrumentation, and protection systems
- Phase D: Final preparations for start-up commissioning
- Phase E: Charge with feedstock and so on. Start up plant and operate
- Phase F: Performance test and plant acceptance

Systematic Commissioning Planning

Systematic and detailed planning begins well in advance of start-up. Contract specifications provide details about start-up, testing, and commissioning. Plant handover, by process system, allows the progressive precommissioning and commissioning of the plant and offers significant schedule advantages. It does, however, require increased planning effort. In addition, a comprehensive safe system of work has to be set up to allow shutdown activities to proceed safely, while precommissioning and commissioning work proceeds in parallel. The phases from mechanical completion to commissioning are as follow (Figure 16.1).

Mechanical completion: Mechanical completion is the term used to cover the phase between equipment installation and the start of process commissioning. From the contractual point of view, mechanical completion occurs between "completion of shutdown" and "acceptance." It may include some specified performance tests and usually refers to individual components of a plant rather than the total plant.

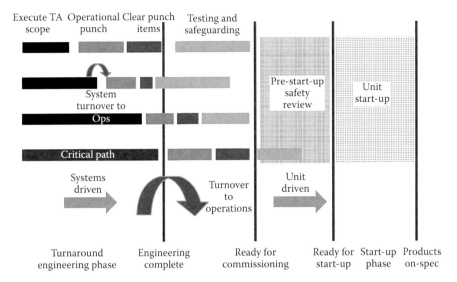

FIGURE 16.1
(See color insert.)

Precommissioning: Precommissioning activities start when the plant, or system, achieves mechanical completion. Frequently, however, precommissioning activities overlap mechanical completion activities and for this reason the plant should be separated into easily manageable system packages, where each system package will be pre-commissioned as a whole and isolations at the boundaries of the system package will be maintained until the completion of precommissioning activities. precommissioning activities include checking for design conformity; checking the status of electrical, mechanical, and instrument installations; running-in of equipment; flushing and cleaning activities; and drying.

Unit ready for start-up: This is the point where the unit is ready to establish process conditions with the intent of making products.

Start-up: Start-up is the point where process fluids and conditions are established with the intent of making products.

Unit ready for commissioning: This is the point where all utilities are commissioned and operational and the unit is ready to accept the introduction of hydrocarbons.

Commissioning: Commissioning is the phase where design process fluids are introduced to the process. Note that for hydrocarbon systems the commissioning activities commence with inserting operations. Commissioning activities normally consist of activities associated with running or operating the plant and include operating adjustments necessary for satisfactory operation of the plant or part thereof. Also included are "functional checks," which are methods used to prove that an item of mechanical equipment or control system functions correctly.

To enable a smooth transition from plant shutdown to commissioning it is required to complete the shutdown work in small packages (known as systems).

System–A section of the plant or facility that can be pre-commissioned and commissioned independently, but in parallel, with other sections of the plant or facility under shutdown.

This enables precommissioning work to progress from an earlier point in the schedule, thus reducing the final precommissioning peak workload substantially. However, systemized turnover is more complex to coordinate and manage due to

- Increase in boundaries between the shutdown stage and commissioning stage
- Increase in parallel activities during the final stages of shutdown

The size and content of a system changes for each shut down job; however, it is essential to take the following points into consideration when defining a system's size and content.

- Units separated into a large number of systems have high manpower expenditure for the precommissioning phase.
- Units separated into a low number of systems have an uneven workload pattern throughout the precommissioning phase.
- Hydrotest boundaries should coincide with system boundaries, where possible.

Commissioning Plan and Critical Path Network

A network is developed that indexes and identifies all start-up activities. Parallel and consecutive activities will be identified. Each activity will have an expected duration and will logically link into the next related activity. Each activity will have a summary sheet in checklist format to identify the necessary actions to be undertaken to achieve the completion of the activity. A strategy will be developed to achieve the following requirements within the agreed contractual milestones.

Mechanical Completion

The objective *of mechanical completion* is to prove that an installed plant component is suitable for commissioning. This phase includes

- Checking that equipment is installed correctly
- Proving that the basic components of equipment operate mechanically as specified, or at least acceptably, for commissioning

- Demonstrating that instruments and control equipment work
- Proving to the commissioning team that components are suitable for precommissioning

Checks are made against the specification and vendors' and contractors' drawings both for equipment details and system completeness.

Process safety and operability are checked—for example, orientation of relief valves, "fail-safe" systems, orientation of nonreturn valves, access to valves, and adequate provision of drains and vents. The plant is finally checked for cleanliness, removal of shutdown debris, and so on.

Physical safety aspects, completeness of safety fittings, handrails, means of access, escape routes, emergency showers and eyebaths, fire extinguishers, and so on are also checked.

Paperwork is required for handover from the shutdown team to the commissioning team—to speed up commissioning it is advantageous for the commissioning team to accept the plant from the shutdown team in sections, so that plant testing and checking can proceed as soon as possible.

Before pressure testing, a functional inspection of each system is carried out to ensure that it has been installed correctly—for example, correct valve types, control valves the right way round, correct instruments fitted, and so on. All vulnerable equipment such as pumps and control valves are fitted with temporary strainers. The system is then blanked off at each end, filled with water, and pumped up to test pressure. The line/vessel test pressures must be compatible. Relief valve set pressures are tested on-site and tagged before installation. If necessary, relief valves are blanked during pressure testing of lines.

The systems should also be checked from an operating viewpoint for proper location and orientation of valves, vents, drains, steam tracing and traps, sample connections, etc. Punch lists of deviations, installation errors, missing items, and so on shall be prepared and given to the appropriate shutdown department for correction or completion. Also included in this inspection and check-out would be vessel internals—this includes going through each tower and vessel to check the installation of trays, baffles, demisters, spargers, etc., and verifying that the dimensions of all weirs, down comers, distributors, thermo wells, etc. are within specification and are properly oriented, as well assuring the internal cleanliness of each vessel prior to final closure.

Mechanical Integrity

The goal of a mechanical integrity program is to ensure that all process plant instrumentation, equipment, and systems function as intended to prevent the release of dangerous materials and ensure equipment reliability. An effective mechanical integrity program incorporates planned inspections, tests, and preventive and predictive maintenance, as opposed to breakdown maintenance. This section examines the aspects of mechanical integrity causally related to the incident.

- Nonoperating adjustments: All nonoperating adjustments, packing of pumps and valves, and cold alignment of equipment performed by the contractors should be spot checked, witnessed, or periodically inspected to ensure proper records are being kept and recorded for turnover.
- Piping: Checklists should be prepared on each hydrostatic test system before the system is released for hydrostatic testing. Completion of this punch list by the shutdown contractor would be required before the test system is released for testing. Post-hydro punch lists should be prepared and followed through to ensure that the systems are fully completed and ready for precommissioning.
- Electrical installation: Each electrical circuit and motor installation and their electrical polarity check and insulation resistance measurement should be demonstrated by the electrical subcontractor to the commissioning subcontractor team.
- Instrument installation: There should be a review of instrument installation for correctness according to the instrument drawings and specification. The commissioning team should witness and/or approve all instrument calibration and check out procedures or work performed by the instrument subcontractor.
- System pressure testing: Although the shutdown contractor is responsible for the actual pressure testing of each system prior to mechanical completion, the commissioning team should review the test program and the methods used to ensure satisfactory tightness.

Punch Listing

Immediately prior to the hydrotest of each system, a punch list will be performed by an integrated punch list team consisting of a shutdown contractor, a site engineer, and production engineers. Punch listing will be undertaken on a system basis. An initial punch list will be undertaken immediately prior to system handover for precommissioning purpose. Punch list items generated at this stage will be incorporated into the master punch list system. A final punch list will be undertaken by the system immediately prior to the commencement of commissioning activities. This will be a less stringent punch list aimed at checking the system before introducing hydrocarbons. Utilization of the precommissioning check sheets will assist in completing a full range of required checks.

Detailed Precommissioning Execution Plan

Precommissioning refers to preparing the plant for the introduction of process materials and its main object is to eliminate any problems which might arise at later and more critical stages of plant operation.

During precommissioning the following activities are carried out:

- Precommissioning activities overlap mechanical completion
- The plant is segregated into manageable systems/subsystems/flushing loops
- Checking of design conformity
- Checking status of electrical, mechanical, and instrument installations
- No load/load run of rotary equipment
- Flushing, cleaning, drying, etc.
- Loop checking should also start after flushing

A plan is developed that identifies all of the major precommissioning activities by the system. The plan will be developed in reverse and will clearly identify "system ready for precommissioning" milestones. Commencement of the system turnover from shutdown to commissioning should ideally take place at approximately 70% erection completion. From this point, handover of systems should be made on a regular basis. Safety is impaired when there is a period of low activity followed by a period of high activity to achieve target precommissioning progress.

A method of precommissioning progress monitoring will be developed to allow progress to be monitored by:

- Overall precommissioning completion for the shutdown
- Precommissioning completion by the process unit
- Precommissioning completion by the system
- Discipline completion by the system

System Cleaning

This includes directing the cleaning of systems either by flushing, blowing, or chemical method by issuing directives for the necessary temporary connections, utilities, and proper lineup of the system to be cleaned. When the services of a chemical cleaning subcontractor are required, the commissioning team will monitor his or her performance to ensure a satisfactory cleaning job. Before each system is closed for operating, the lines and equipment would be inspected for cleanliness. Cleaning progress will be recorded on an A3 set of P&IDs.

Equipment Run-In

This includes planning and coordinating vendor representatives for the running in of the equipment and obtaining the necessary utilities to accomplish the run-in for equipment such as blowers, pumps, compressors, motors, turbines, and other machinery or mechanical equipment without process fluids before start-up.

Temporary works: The nature and extent of any temporary works should be reviewed during the early stages as these can have a major influence on the manner and approach to commissioning

The following should be considered carefully:

- Features that will aid commissioning—for example, additional branches to allow the hook-up of commissioning equipment such as pumps, installments
- Temporary equipment—for example, additional pumps, bypass lines, storage tanks, strainers, and blanking plates
- Temporary supplies of consumables—for example, portable gas supplies and feedstock (both primary and intermediates)

High-pressure steam lines to turbines must be meticulously cleaned by repeated blowing through to atmosphere at high velocity against target plates to achieve the standard of cleanliness required before coupling to the driven equipment; usually this is carried out in the presence of the turbine vendor's commissioning engineer. Chemical cleaning of boiler systems, compressor connections, and any special plant pretreatment is carried out as necessary.

Air or nitrogen pressure testing up to normal working pressure is carried out to ensure system tightness. Specialist electrical, mechanical, and instrument commissioning personnel concentrate on their specific tasks throughout this period. Machinery is run lightly and functional testing of instrumentation and control loops is carried out after calibration and setting up. Trip systems and other safety devices are carefully checked out in conjunction with process commissioning personnel.

The equipment vendor's commissioning personnel are called forward as appropriate, care being taken with regard to timing and providing adequate notice.

Throughout the precommissioning period, it is highly desirable that operating and maintenance personnel of the owner participate actively in all that goes on.

The plant must be cleaned up to the commissioning manager's satisfaction, which usually means that all rubbish, trip hazards, nonessential scaffolding, and other materials (especially flammable or toxic) should be cleared away;

Precommissioning of Field Instrumentation

During commissioning, problems are often encountered with the instrumentation. Every effort must be made to ensure that the instruments are functioning accurately and correctly before start-up and have not been damaged during commissioning.

Ideally the field instrumentation and process connections—for example, impulse lines for dp cells—should have been properly installed prior to commissioning as should the infrastructure such as power and air supplies, conduits/trunking/cable trays.

The emphasis in precommissioning is on checking. An essential first step in the precommissioning of instrumentation is a visual inspection of the installation such as its mounting, associated pipework, and wiring, and to check that the workmanship is of an acceptable standard.

Control System Commissioning

The principal objective of the commissioning process is to identify faulty installation and/or operation of the hardware and mistakes in the software. Its scope therefore embraces inaccurate calibrations, incorrectly wired i/o channels, faulty configuration, incomplete operator displays, and illogical sequence flow.

The basic strategy for commissioning instrumentation and control systems is the systematic functional testing of all the elements and subsystems against the functional specification.

Electrical System

All earth continuity is checked and an acceptable level of grounding resistance of the structure and vessels achieved. Wiring is checked for continuity and insulation resistance. Finally, motors must be checked for correct direction of rotation.

Precommissioning and Commissioning Manuals

Precommissioning and commissioning manuals will be supplied as a collection of precommissioning procedures; these procedures are operating instructions for the activities to be carried out during the precommissioning and commissioning activities.

The main list of the procedures to be carried out is

- Mechanical preparation
- Chemical cleaning instructions
- Physical cleaning instructions
- Mechanical restoration

- Machinery run-in
- Tightness test
- Electrical testing/functional tests/energizing
- Instruments calibration and functional test
- Loading of chemicals
- Loading of catalyst
- Heaters drying

What to Do in Checklisting

- To be done as per system/subsystems
- System is as per P&ID
- Identifying skewness of flanges
- Short/long bolting
- Use of proper gasket—color codes
- Support systems
- Location of drains/vents
- Columns/Vessel—internals as per drawing
- Positioning of instruments
- Valve/CV installations

Commissioning

During Commissioning and Start-Up Period

- Find out the commissioning critical path
- Find out the start-up procedure
- Licensor/vendor support
- Conduct and supervise all activities during the introduction of feed-stock to the units
- Stabilize the unit
- Conduct performance test run

Process Commissioning

Process commissioning begins when the precommissioning tests have been completed, and faults corrected to the satisfaction of the commissioning manager. This is when plant, equipment, systems, or subsystems are put into operation for their normal duty. As discrete systems or subsystems become available after completion of the precommissioning tests, and faults have been corrected, it is often possible to commission them on process fluid.

Ideally, the commissioning team should consist of engineers with relevant experience who have previously worked together and have sufficient depth of experience in all aspects of plant precommissioning and commissioning. A successful team may well include a balance between those of theoretical and practical ability.

Manpower Deployment

- Multidisciplinary core team including team leader
- Blend of experienced and new operating personnel including mechanical, electrical, instrumentation, process, etc.
- Deploying few senior members
- Team leader should be able to command the resources

Commissioning and Start-Up Period

Once the unit is pre-commissioned, the commissioning team will conduct the following main activities:

- Purging
- Leak test/tightness test
- Spring hanger check

To begin initial start-up operations assuring that every activity is accomplished smoothly and safely, the commissioning team should take care of troubleshooting any operating or process problems, should they arise, and should investigate any operating deviations as they occur.

Minimum Facilities before Start-Up

- Implementation of work permit system original
- Firefighting facilities including positioning of extinguishers etc.
- Clearly defined emergency exits, with display boards
- Removal of debris/temporary structures and power connections
- Testing paging systems and availability of walkie-talkies
- Testing oily waste sewer/contaminated rain water sewer system with an effluent treatment plant in operation
- Availability of adequate personal protective equipment/first aid boxes
- Judicious implementation of isolation scheme
- Properly stenciled utility points with connecting hoses/lancers
- Unlocking of springs/supports of all headers
- All gas/hydrocarbon detectors are tested and kept on
- Electrostatic-sensitive devices are tested, documented, and kept activated
- Flare system is in position

Pre-Start-Up Safety Review (PSSR)

If the PSSR, which is a formal safety review by a technical team led by the operations superintendent, had been implemented, a technical team would have verified the adequacy of all safety systems and equipment, including procedures and training, process safety information, alarms and equipment functionality, and instrument testing and calibration. The PSSR requires a sign-off that all nonessential personnel have been removed from the unit and neighboring units and that the operations crew have reviewed the start-up procedure.

Customized Testing and Acceptance Methodologies

The testing and acceptance procedures are the most critical and important component of any plant commissioning process. How items will be tested and when these tests are considered completed determines a successful transition; tests and acceptance methods define what is checked and when it is acceptable. Equally important, the plant operations group develops a level of confidence with the work and new systems during the test and acceptance stage. During the project, the testing program is broken down into two major stages as detailed below.

Operational readiness test (ORT): Testing by the contractor of the installed system to verify operational integrity of the system under operational conditions without being placed in actual operation. Steps completed during this test include aligning and leveling equipment, wiring and loop checks, calibrating all instruments, etc. During this function test, representatives of all major components for the equipment are on site for timely identification and resolution of equipment issues.

Performance acceptance test (PAT): The performance acceptance test verifies the operation of the system under day-to-day usage, after installation and start-up are completed and the system is fully functional. Any system deficiencies are noted, and the system downtime is assessed. This test is performed by the contractor, with support provided by the engineer and plant operations group.

Sequence of Commissioning

Utilities

- Electricity, including emergency generators.
- Water and its treatment systems; compressed air for instruments, process, and breathing.
- Demineralized water.
- Natural gas receiving station.
- Firefighting water.
- Steam boilers and economizers.
- Condensate.
- Nitrogen or other inert gas from cylinders or a production unit. These must be regarded as process chemicals and blanked off from process equipment until otherwise authorized.
- Drains and effluent treatment systems.

The commissioning manager's main concern at start-up is to achieve stable process conditions as quickly as possible and then to be confident of achieving a reasonable running time without breakdown.

- Warm-up routines complete where necessary—for example, steam or electric tracing on feed and product lines and process equipment
- Establishing inert atmospheres

- Ancillary systems operating normally
- Cooling water and emergency systems for bringing potentially hazardous reactions under control are immediately operational if required
- Stabilize the unit—the commissioning team will work with process group on how to bring the units on stream and up to design rate. Control variables and flow rates will be specified to stabilize the unit at design rate in preparation for the acceptance test and during the performance test.

List of Items in Checklist before Start-Up (*After M&I Shutdown*)

1. General points:
   - Complete all jobs in each individual equipment.
   - Complete de-blinding/blinding jobs as per standard master list.
   - Close all notifications issued for various jobs during shutdown.
   - While boxing up any equipment, a production representative must be physically present at the site to ensure that no person is left inside.
   - Clean the unit area thoroughly.
   - Remove all temporary lines, cables, and facilities.
   - Complete insulation jobs—both laying mattress and sheeting.
   - Remove soaked insulation completely.
   - Check for short bolting in any equipment and rectify the same.
   - Ensure availability of additional utilities, if required.
   - Ensure availability of feed.
   - Keep fire tender as standby as per requirement of the unit.
   - Ensure availability of experienced, adequate manpower both in the field and panel areas.
   - Prepare utilities—drain condensate from all steam headers, sub-headers up to the points of consumption to make dry steam/avoid hammering. Drain water from service air and instrument air lines.
   - Ensure availability of chemicals required for start-up and normal running of the unit.
   - Remove all scaffolding materials, project materials, civil materials (insulation, refractory, etc.), unused materials like pipes, plates, equipment, scrap materials, etc. from the unit area.
   - Commission utilities after thorough checkup for complete draining of condensate through traps and their bypasses.

- Energize motors after taking over from equipment maintenance and thoroughly checking both in the field and panel areas.
- Keep vaporizer system in readiness to meet the start-up (gas backup) fuel requirement through lpg, if required.
- Remove air from the equipment and piping by water/steam as per practice and finally with hydrocarbon.
- Drain held-up water from the equipment, pipelines. Vent adequate steam for elimination of air from the equipment completely.
- Test all equipment thoroughly for any leakage etc. and take corrective measures immediately.
- Carry out pressure testing of the equipment with steam/air/nitrogen as per requirement.
- Follow refractory dry out cycle as per procedure.
- Follow unit start-up procedure as per normal practice.
- Do not allow any hydrocarbon to escape into the surroundings. All draining should be through a closed blow down (CBD) system/OWS as per normal practice.
- Follow equipment start-up procedure religiously.
- Stop hot jobs at all locations in the unit and nearby vulnerable areas.
- Use standards such as work permit system, handling of air–hydrocarbon mixtures and pyrophoric substances, etc.
- Check alarms and securities prior to start-up of the plant.
- Drain condensate from the steam turbine enclosures, supply line, etc. periodically.
- Follow normal procedure for starting up hot pumps like warm up etc.
- Drain water thoroughly from the feed tanks, supply lines, fuel supply system, etc. Special care should be taken for the lines and equipments which were hydrotested.
- Simulation and checking of interlocks jointly by production and instrumentation maintenance.

2. Furnace:

- Ensure removal of all foreign materials including mucks etc. from all the platforms of the heater.
- Check all the ignitors for their workability.
- Eliminate chances of liquid carryover with fuel gas completely prior to burning in the heaters by draining periodically to close blow down (CBD).

- Ensure proper heating of liquid fuel prior to feeding to the burners. Establish long circulation of liquid fuel in the supply system at desired pressure and temperature to avoid spillage through burners. Check proper opening of back pressure control valve in the liquid fuel supply system.
- Ensure draining/displacement of total water from the heater coils and establish circulation in all the tubes of the heater. No pocket of water should be allowed.
- Follow drying cycle during start-up for long life of the refractory lining. Increase temperature of heater as prescribed for minimizing thermal shock.
- After every unsuccessful firing of the heater or flare, flush out the held-up material thoroughly with steam or air (as the case may be). This will reduce chances of backfire during lighting up activity.
- Check burner alignment, refractory condition, thorough cleaning of burners, etc. prior to start-up of heater.
- Try to maintain uniform distribution of air through air registers. Check air pressure both at plenum chamber and at individual burner box.
- Check operation of shutoff valves (SOVs) in the fuel lines to the furnace. SOVs are required to be reset once they operate to avoid backfiring/accident.
- Maintain the desired "delta p" across fuel oil and atomizing steam supply to burners for efficient burning.
- Keep a constant watch on the tube metal temperature of the furnace particularly during start-up for immediate detection of any abnormality.
- Switch over to balance draft, wherever available, at the earliest opportunity for proper functioning of heater and energy conservation.

3. Columns:
   - Ensure thorough cleaning of all trays, packing beds, downcomers, draw off and reflux nozzles, distributors, etc.
   - All nozzles of the distributors should be thoroughly cleaned.
   - Ensure removal of all foreign materials including mucks etc. from all the platforms of the column.
   - Ensure replacement of soaked insulation in all the lines connected to the column.
   - Ensure thorough cleaning of the column bottom including coke trap, vortex breaker, etc.

- Ensure thorough cleaning of suction strainers for all pumps including bottom pumps.
- Ensure complete air removal during the steaming stage and gas backup simultaneously with the closing of the column vent. Special care should be taken to avoid formation of vacuum during gas backup operation.
- Detect any leakage from the system from the fuel gas smell during backup operation and rectify the same immediately.
- Check all chemical dosing lines to the column for removal of air, gas backup, and finally filling with the desired chemical.
- Check for completion of all insulation jobs, blinding/de-blinding of nozzles, removal of scaffolding, etc.
- Establish cold circulation and check for any leakage etc. and arrest it immediately.

4. Other areas:

- Establish circulation and drain water from all low-point drains.
- Ensure closing of all valves at the low-point drains (LPDs) and high-point drains (HPDs) and cap them properly.
- Carry out hot bolting at desired temperature levels.
- Ensure all pressure safety valves (PSVs) are in position and properly lined up.
- Keep a watch on all vulnerable points for smoking, leakage, and fire, and rectify the same immediately.
- Keep manpower ready for hot bolting with non-sparking tools and tackles.
- Check holding of vacuum in vacuum column.
- Check all the stop switches of various equipments from the panel.
- All portable monitors like explosimeters, gas detectors, carbon monoxide (CO) monitors, etc. should be kept ready for use during start-up.
- All emergency lights in the field and panel areas are to be tested and adequate numbers of safety torches are to be kept ready.
- Talk-back system and also adequate numbers of walkie-talkies to be kept as standby.
- Slop to be diverted to tank after proper cooling. Hot slop should not be diverted to avoid accidents.
- Air should not be pulled through the catalyst bed when the catalyst beds are hot.
- Maintain a positive nitrogen flow on both sides of any flanges that are to be opened for blinding in a hydrotreatment unit.

- For the protection of the heater tubes, they must be neutralized as per requirement.
- Any austenitic stainless-steel equipment is to be opened to the atmosphere after thorough neutralization of the residual polythionic acid with suitable passivation (soda ash) solution.
- The fractionation towers having packings should be washed thoroughly with potassium permanganate water before admitting air to prevent ignition of iron sulfide scale.
- Special care must be taken during shutdown to avoid personnel entering an atmosphere contaminated with hydrogen sulfide or tested to insure the safety of entry.

Pre-Start-Up of Furnace

No.	Description	Status	Date/Time	Checked by
1	Are there any foreign materials inside the furnace and oil-soaked insulation, combustible material, etc. in the furnace premises?			
2	Has the furnace been boxed up? Are all peepholes, explosion doors, manholes, etc. closed?			
3	Are the flue gas main stack damper and (individual furnace dampers) in open position?			
4	Are induced draft (ID) and forced draft (FD) fans energized and ready for start-up?			
5	Have all the burners been handed over after maintenance and are they boxed up?			
6	Is the fuel gas control valve boxed up in the right direction?			
7	Have the instrument connections to the fuel gas control valve been given?			
8	Has the stroke check for the fuel gas control valve (FG C/V) been done?			
9	Is the fuel oil control valve boxed up in the right direction?			
10	Have the instrument connections to the fuel oil control valve been given?			
11	Has the stroke check for the Fuel oil C/V been done?			
12	Is the atomizing steam control valve boxed up in the right direction?			
13	Have the instrument connections to the atomizing steam C/V been given?			
14	Has the stroke check for the atomizing steam C/V been done?			

Pre-Start-Up of Furnace (continued)

No.	Description	Status	Date/Time	Checked by
15	Has the furnace interlock been checked?			
16	Has the damper operation been checked?			
17	Has the fuel oil isolation valve at battery limit been de-blinded?			
18	Are the individual isolation block valves of each burner and its corresponding pilot isolated?			
19	Are the fuel gas, fuel oil, and pilot gas headers up to pressure indication control (PICs) steamed for air removal?			
20	Have condensates been drained from furnace oil, fuel gas, and pilot gas headers?			
21	Are the individual burner air resistors closed?			
22	Is the igniter in working condition?			
23	Is the FG strainer in line and is it clean?			
24	Has the box-purging line been de-blinded?			
25	Has the fuel oil line emergency shutdown logic been bypassed?			
26	Has the atomizing line been de-blinded?			
27	Has the FG line been de-blinded?			
28	Are the furnace's dampers wide open?			
29	Has box purging been done?			
30	Is the steam tracer to FG, FO, and pilot gas charged?			
31	Are the fuel oil low-pressure trip switch and other interlocks bypassed to start with?			
32	Are the isolation valves of the FO manifold open?			
33	Has FO circulation in the furnace ring been established and continued till the system is heated up?			
34	Has FD fan been started?			
35	Is the FD fan discharge flow OK?			
36	Are the isolation block valves in the FG and pilot gas manifold open?			
37	Is the furnace draft gauge showing healthy?			
38	Is the fuel gas/pilot gas low-pressure trip switch bypassed and fuel oil low-flow trip bypassed to start with?			
39	Have all the thermocouples, for example, skin, box, coil, and outlet, been checked and made OK by instrument maintenance?			
40	Ensure that fuel gas/fuel oil, pass flows, (turbulizing steam), etc. control systems are checked by instrument maintenance and found OK for operation.			

Equipment Operating Procedures

Centrifugal Pumps

Pumps in hydrocarbon service are generally provided with mechanical seals. This seal ensures complete prevention of leakage and requires minimum attention during the service life. The sealing is achieved by two mating faces—one rotating with the shaft and the other stationary in the casing. A film of liquid between the two faces is necessary to act as lubricant and coolant. Depending on the type of product handled by the pumps, this liquid called seal-flushing liquid is supplied either from the discharge of the same pump or from an external source. In this unit, the seal-flushing liquid is provided from the respective pump discharge itself. De-pressurizing and draining of the pump casing is done through the closed blowdown drum (CBD) header. The stabilizer reflux pump vents are connected to flare for de-pressurizing.

The procedures for start-up, shutdown, and troubleshooting of centrifugal pump are described in the following sections. In addition, the vendor's operating instruction should be referred to for further details.

Preliminary Checks for Start-Up

1. If starting for the first time after completion of maintenance work, get a mechanical man to see if the pump is in order.
2. Get an electrician to check the pump motor, the switch, and the circuit breaker in the substation if starting up for the first time after maintenance. Get terminal boxes connected and motor meggered. Then give power to the motor.
3. Check that the motor of the pump is properly earthed to prevent electrical shock.
4. Ensure that protective covers are placed over exposed rotary parts of the pump such as pump coupling etc.
5. Check that bearing housing has an adequate oil level of the proper grade. Change oil if it is dirty/contaminated.
6. Confirm whether cooling water lines are connected properly and there is sufficient flow of water and drive out air from casing through the vent.
7. Open water/steam quench to the mechanical seal as per provision.
8. Rotate the pump shaft by hand to ensure that it is free and coupling is secured.
9. Ensure that all blinds have been removed from the suction and discharge line of the pump. Proper pressure gauge should be available on the discharge line.

10. Check whether all bleeder valves have been closed on suction and discharge line of the pump. Now open the suction valve. CBD valves are to be kept closed and the flange blinded.

11. Start the pump and check the direction of rotation. Rectify the direction of rotation if it is not right.

Note: If the pump is in hot oil service, it should be gradually warmed up to a temperature close to that of handled fluid. This is absolutely essential, otherwise the pump casing, impeller, etc. may get damaged due to thermal shock. Hot pumps are normally provided with a bypass across the check valve to keep the idle pump hot.

The following steps are to be followed to prime a hot pump:

1. Crack open the bypass nonreturn valve (NRV). Do not open the bypass NRV too much, else the pump will rotate in the reverse direction.

2. The pump is to be slowly heated to an operating temperature (not more than 2°C/minute).

Pump Start-Up

1. Open all valves on the discharge line except the one closest to the pump. Check line for open bleeders and missing plugs.

2. Start pump by pushing the starter button.

3. Once the motor is started, check for discharge pressure; if it is close to shut-off pressure and whenever amps can be seen, check that amps count after initial torque has come down, then slowly open the discharge valve and check for discharge temperature and flow. Discharge valve opening should not be very slow to avoid the pump heating up.

4. Check the warming up of the bearings and stuffing boxes. The temperature of the bearing should normally be within 60°C.

5. Check whether the motor is unusually hot; in such cases, stop the pump.

6. In case of vibrations, stop the pump and report to a mechanical representative for necessary action, that is, alignment checking etc.

Normal Check-Up of Pumps

1. Check for any unusual sound.

2. See that mechanical seals are not leaking and pump-cooling water and seal-flushing liquid are flowing properly.

3. Ensure that bearing housing has sufficient lubricating oil.

4. Note whether pump is discharging at normal pressure.

5. Confirm that motor is not excessively hot and bearings are not making any sound, and check that the pump load is within limits.

6. See that motor earthing is not damaged.

Positive Displacement Pumps

Start-Up

1. Check whether all mechanical jobs are over.
2. Ensure that the gearbox has sufficient lube oil of proper grade. Change oil if it is dirty/contaminated.
3. Confirm the presence of a strainer in the suction.
4. Check for proper lining up including pressure safety valve in the discharge. Open wide the suction valve.
5. Rotate the motor shaft by hand and see that it is free and that coupling is secured. The coupling guard should be in position.
6. Connect the terminal box and megger, and energize the motor. Open the discharge valve, start the motor, and check the direction of rotation. Stop and rectify if the direction of rotation is wrong.
7. Adjust the pump stroke and run the pump at different settings. Watch the discharge pressure and check the rate of pumping using a calibration pot if provided. In case the pump is provided with a pressure controller, adjust it to obtain the required discharge pressure.
8. Care should be taken to avoid dry running of pump and backflow of liquid. Bleed if necessary to expel vapor/air.
9. Check for unusual noise, vibrations, and rises in temperature of both motor and gears.
10. Check the discharge safety valve is not passing to suction.

Shutdown

1. Stop the pump.
2. Close the suction and discharge valves.
3. Drain the liquid if maintenance jobs are to be carried out on the pump. De-energize and disconnect electrically. For major maintenance, blind suction and discharge.

Heat Exchangers

Taking Over of a Heat Exchanger after Mechanical Repairs

Heat exchangers are hydraulically tested and certified as fit for operation before being handed over to the operating staff.

The operating staff should check the following while taking over a heat exchanger:

1. All nuts and bolts are in position and tight.
2. All drains are capped/blinded.

3. Thermo-wells, if any, are in position.

4. Thermal insulation, if any, is provided and blinds from the inlet and outlet of the shell and tube side are removed.

Commissioning of a Heat Exchanger

While commissioning a heat exchanger, the following points must always be remembered:

1. The cold side must be commissioned first.

2. Thermal shocks must be avoided.

3. Air from the exchanger must be vented and that can be done as follows:

   a. Displace air from the exchangers with flushing oil through the top vent to the oily water sewerage (OWS) with a hose connection till flushing oil comes in full bore.

   b. Air from the coolers/condensers is to be displaced with water from the top hose connection till water from the top comes in full bore. The water hose can be connected at the bottom of the cooler/condenser. Then back up with product/gas and displace/drain water from the bottom.

Commission cold side as follows:

1. Crack open the outlet valve on the cold fluid side and then slowly open the outlet valve fully.

2. Gradually open the inlet valve on the cold side till it is wide open. Also open the hot side outlet valve slowly till it is wide open.

3. Start slowly closing the bypass valve on the cold side. The last 4–5 threads of the bypass valve of the cold side are to be closed very carefully.

After commissioning the cold side, establish the flow through the hot side as follows:

1. Gradually open the inlet valve till it is wide open.

2. Start slowly closing the bypass valve on the hot side. It is essential to operate the valves gradually, otherwise the exchanger will start leaking because of thermal shock. The last 4–5 threads of the bypass valve are to be closed very slowly to avoid upset/thermal shock.

3. While venting air from the exchanger during filling, care should be taken to avoid spillage of hot oil. A hose should be connected from the vent so that material is safely routed to the OWS.

4. Get CBD blinded.

Flange joint leakage can occur because there is no department or person responsible for providing support or addressing bolting and leakage problems, or there is no technical "expert" to address bolting or bolted-gasketed joint issues. In general, the understanding by plant staff of various leak mechanisms, required gasket stresses, surface finish, bolt preload, and impact of service conditions is weak to nonexistent. Other than "jump-up" repairs, the leakage situation is not getting the attention that is needed to address and correct the root causes of leaks.

There is no keen awareness of the overall leakage situation. This results in inadequate direction and expectation from plant management as to the extent of leakage (if any) that is acceptable.

The result of these deficiencies is that many of the leaks return. At some point, plant management became aware that action must be taken. The first step was to assign a member of the engineering department the responsibility for resolving all leakage and bolting concerns.

17

Postshutdown Review

We judge ourselves by what we feel capable of doing, while others judge us by what we have already done.

Henry Wadsworth Longfellow, American poet

The completion of one turnaround is the next shutdown's starting point. The stage covers demobilization of contractors, lay-down area cleanup, disposal of excess material, documentation and updating the turnaround historical database, cost reports, and most importantly lessons learned that could be carried forward to the next turnaround. Executing this stage in a timely manner with a quality result will depend on data collection effectiveness during the execution stage.

It is important to monitor both the administrative and contract closure of the shutdown. It is imperative that communication between these two sides of the closeout takes place continuously throughout this final stage.

Contract Closure

Contract closure follows with making sure the contractor or vendor receives final payment as well as documentation and evaluation of how services were provided throughout the contract. It is not uncommon to detail both the positives and negatives which occurred during the duration of the contract. Violations may be documented as well as awards given for good work.

It also determines if the work described in the contract was completed accurately and satisfactorily. Contracts may have specific terms or conditions for completion and closeout. You should be aware of these terms or conditions so that shutdown closure isn't held up because you missed an important detail.

Administrative Closure

Administrative closure takes care of all the paperwork associated with the shutdown while making sure that all documentation is updated, minutes are clear and concise, and all analyses associated with the shutdown have been detailed for historical purposes.

During closure, it is very important for the shutdown manager and his or her team to make sure the documentation is complete. This documentation may be used for future shutdown.

Administrative closure includes the following:

- Documentation of the shutdown work
- Confirmation that the product is in alignment with requirements and specifications
- Analysis of shutdown success or failure
- Analysis of the effectiveness of the management process
- Lesson learned documentation

Documenting the Shutdown

Documentation always seems to be the most difficult part of the shutdown to complete. There is little glamour doing documentation. That does not diminish its importance, however. There are at least five reasons why we need to do documentation:

Reference for future changes in deliverables: Even though the shutdown work is complete, there will be further changes that warrant follow-up shutdowns. By using the deliverables, the customer will identify improvement opportunities, features to be added, and functions to be modified. The documentation of the shutdown just completed is the foundation for the follow-up shutdowns.

Historical record for estimating duration and cost on future shutdowns, activities, and tasks: Completed shutdowns are a terrific source of information for future shutdowns, but only if data and other documentation from them are archived so that they can be retrieved and used. Estimated and actual duration and cost for each activity on completed shutdowns are particularly valuable for estimating these variables on future shutdowns.

Training resource for new shutdown managers: History is a great teacher, and nowhere is that more significant than on completed shutdowns. Such items as how the work breakdown structure (WBS) architecture was determined, how change requests were analyzed and decisions reached, problem identification, analysis and resolution situations, and a variety of other experiences are invaluable lessons for the newly appointed shutdown manager.

Input for further training and development of the shutdown team: As a reference, shutdown documentation can help the shutdown team deal with situations that arise in the current shutdown. How a similar problem or change request was handled in the past is an excellent example.

Input for performance evaluation by the functional managers of the shutdown team members: In many organizations, shutdown documentation can be used as an input to the performance evaluations of the shutdown manager and team members.

Inspection outlook report: The inspection outlook report is prepared by the inspection department and is a very comprehensive document covering various aspects of the equipment health and expected life. The report should be released within 6 months of the turnaround covering the following details:

- It should highlight the jobs done during the turnaround equipment-wise.
- It should mention jobs which could not be done due to some constraints.
- It should note observations recorded for critical equipment post turnaround, for example, skin temperatures for refractory lined surfaces, heat transfer and pressure drop of heat exchangers, etc.
- It should describe detailed recommendations equipment-wise, giving repairs/replacement anticipation in the short and long terms.

Postevaluation Report

The post-execution audit is an evaluation of the goals and activity achievement as measured against the plan, budget, time deadlines, quality of deliverables, and specifications, and submitted to the department head for review.

1. Was the goal achieved?
   a. Does it do what the team said it would do?
   b. Does it do what the management said it would do?

   The job was justified based on a goal to be achieved. It either was or it wasn't, and an answer to that question must be provided in the audit. The question can be asked and answered from two different perspectives. The provider may have suggested a solution for which certain results were promised. Did that happen? On the other hand, the requestor may have promised that if the provider would only provide, say, a new or improved system, certain results would occur. Did that happen?

2. Was the work done on time, within budget, and according to specification? In the scope triangle, the constraints on the shutdown are time, cost, and the customer's specification, as well as resource availability and quality. Here we are concerned with whether the specification was met within the budgeted time and cost constraints.

3. Was business value realized? (Check success criteria.) The success criteria were the basis on which the business case for the shutdown

was built and were the primary reason why the shutdown was undertaken. Did we realize that promised value? When the success criteria measure improvement in profit or market share or other bottom-line parameters, we may not be able to answer this question until sometime after the shutdown is completed.

4. What lessons were learned about your shutdown management methodology? Companies that have or are developing a shutdown management methodology will want to use completed shutdowns to assess how well the methodology is working. Different parts of the methodology may work well for certain types of shutdowns or in certain situations, and these should be noted in the audit. These lessons will be valuable in tweaking the methodology or simply noting how to apply the methodology when a given situation arises. This part of the audit might also consider how well the team used the methodology, which is related to, yet different from, how well the methodology worked.

5. What worked? What didn't? The answers to these questions are helpful hints and suggestions for future shutdown managers and teams. The experiences of past shutdown teams are real "diamonds in the rough"; you will want to pass them on to future teams.

Review meeting: The shutdown end should also mark a time for reflection and constructive criticism while the shutdown is still fresh in everyone's mind. Much can be learned by reviewing what went right and what went wrong during the shutdown. Post-turnaround review meetings should be conducted in the following separate groups.

Review with department heads: This meeting is held to take stock of weaknesses/ deficiencies in the systems that came to light during the turnaround and the failures that had to be faced. The aim is to find out "what went wrong" and not "who went wrong" so that required steps can be taken to bring about necessary improvements.

The deliberation in this meeting should be on the following lines:

- Did shutting down the process units take longer than planned. If so, the reasons thereof?
- Did some jobs on the critical path take longer than planned? Did some other jobs take still longer time to complete?
- From the monitoring that was being done during the turnaround, it would be indicated which jobs took longer than the planned time. A dispassionate look at how the job was done can indicate whether the planned time was too little or that actual time taken was more than necessary.
- Did some contractors prove not so good despite their earlier good reputation?

- A good discussion with them individually (keeping the contractual agreement out of view) can bring out the reasons.
- Did quite a few jobs come up at the 11th hour? Did accepting them without proper assessment upset the plan?

Review with contractors: The maintenance head along with other department heads like inspection, finance, and materials holds a meeting with the representatives of all the contractors who participated in carrying out the turnaround work and shares views. The contractors are encouraged to give their views on various procedures or any other points which possibly prevented them to give their best and how the work could have been organized better. The inspection head may express his or her views on the quality of work generally executed during the turnaround and point out how quality could be improved. A feedback can assist in bringing about required improvements in future turnarounds. Various problems faced by the vendor should be discussed so that the same may be overcome to improve the quality, efficiency, and cost.

Review with supervisors/everyone: A post-turnaround meeting should be convened by the working committee head with all supervisors/everyone who had been assigned responsibilities during the shutdown as well as coordinators from all departments to get feedback from them and to share experiences with them. Individuals are encouraged to speak out their mind so that improvements could be made during the next turnaround. As this will include a good number of individuals, this meeting is often best scheduled after regular work hours. Sometimes this review is best conducted off-site so as to minimize interruptions.

The goal of the shutdown review meeting is to provide an open forum where general aspects of the shutdown can be discussed. In addition, it is at this review where specific assignments are made relative to written reports that will become the final report document and file.

Specifically, the shutdown review meeting should cover the following topics that would be of general importance to the entire shutdown.

Safety/Environmental Performance and Procedures

Accidents during turnaround: Safety is accorded a very high rating in a petroleum refinery. Apart from the social angle of accidents, most companies recognize that they cannot attract good talent if their refinery is considered hazardous. The high incidence of accidents can itself be a reason for slowing down execution of work resulting in a delay in completion of the turnaround. The accidents will need to be thoroughly analyzed one by one and reasons carefully identified.

Accidents: All accidents should be discussed and recommendations made to ensure that there are no repeats during future shutdowns.

Fire incident during the turnaround: As stated earlier, a fire incident during a turnaround can disrupt the tempo of work. Even if the incident did not cause an overall delay in the completion of the turnaround, the incident indicates a system failure.

- Near misses: Any near misses should also be discussed in the final report so that such occurrences don't become the accidents of the next shutdown.
- Permitting: Any delays or problems associated with the acquiring of clearances on machinery or entry into lines or vessels should be discussed.

Alternative plans to eliminate such delays or problems should be brainstormed. If such instances were numerous, a study group should be assigned to come up with proposals for future shutdowns.

- Environmental incidents: Discuss any environmental considerations, such as emissions or hazardous waste handling.
- Personnel protection: Identify any deficiencies in the availability or suitability of personal protective equipment. Develop recommendations for future shutdowns.

Contractor Performance

- Training: Identify any additional "right-to-know" requirements or means to minimize such training in future shutdowns by additional barricading or restricted work areas.
- Performance: Identify any contractors that did not meet expectations or performance. Identify whether such nonconformance could be eliminated through tighter contracts, additional pre-job site inspections, etc.

Schedule Compliance

- Schedule breaks: Identify all instances which caused schedule breaks, interruptions, or delays. Brainstorm innovative ways to minimize such disruptions in the future.
- Logistics: Identify deficiencies in logistical areas. Propose other methods for delivery of materials, equipment, or supplies.
- Mobile equipment: Discuss problems associated with mobile equipment needs, scheduling, and placement. Give special attention to alternate lift paths, or routing for heavy lifting equipment.

Work Assignments and Execution

- Resource allocation: Discuss any unresolved problems with resource allocation. Specifically target jobs that were left incomplete at the end of the shutdown.
- Execution reports: Assign report requirements to all who had specific shutdown assignments. A response date should be stipulated for these reports. Usually, 2 weeks is more than adequate. Allowing longer than this will invariably cause some individuals to put it off to the last minute. As time lapses, the actual details of the shutdown will become clouded for such individuals.

Costs

Assign the responsibility of accounting for the shutdown costs. If a common account code or shutdown cost account was established, collecting the costs is made relatively easy. Take into account that costs will continue to come weeks after the shutdown is completed. If a final accounting tally must be made, these remaining costs can be estimated.

It is usually wise to estimate them higher than lower. If the actual costs come in lower than estimated, some department or cost center other than maintenance will be pushing to get the overrun corrected. If the actual costs come in higher than estimated, maintenance will have to exert their authority to have the corrections made.

The other points discussed are

- Shutdown organization including staffing and skills
- Successful risk assessment and mitigation techniques, that is, what risks occurred and what techniques were used to mitigate these risks
- Processes used for change management and quality assurance
- General techniques used for shutdown communication
- General techniques for managing customer expectations
- Short-term success factors and how they were met
- Culture or environment
- Lessons learned (from the lessons learned session)
- Recommendations to future shutdown managers
- Shutdown team recommendations: Throughout the life of the shutdown, there will have been a number of insights and suggestions. This is the place to record them for posterity.
- Techniques used to get results: By way of a summary list, what specific things did you do that helped to get the results?

- Shutdown strengths and weaknesses: Again by way of a summary list, what features, practices, and processes did you use that proved to be strengths or weaknesses? Do you have any advice to pass on to future shutdown teams regarding these strengths/weaknesses?

Minutes

Prior to the shutdown review meeting, assign a capable individual to keep accurate minutes of the meeting, including all discussions and recommendations. As an alternative, the meeting should be taped so that it can be transcribed later.

Shutdown File

While the individual reports are being prepared, the shutdown file can be started. This file should be the repository of all hard copy generated in connection with the shutdown. Specifically, the file should include

- All work orders initiated for the shutdown: Set up subfiles for completed, unfinished, canceled, and new work.
- Contracts: Include originals on all contracts entered into for the shutdown.
- Insurance papers: Include a file for all proof-of-insurance documents.
- Equipment reports: Include original reports prepared for the work actually done during the shutdown.
- Drawings, prints, and sketches: Include copies of all documents, or a listing of where these documents exist.

Final Report

All of the information gathered and assigned at the shutdown review meeting should ultimately end up in a final report. In addition to the specific details recorded at the meeting and the equipment repair reports subsequently prepared, the final report should even include the broad description of the shutdown's major focus.

The final report should be a cogent document that is an accurate description of what was done, what needs to be done, and what it all cost.

Final Report on the Shutdown

A report that documents important results during the effort also needs to be prepared. This report must capture work that needs to be considered in the next shutdown. The preparation of the final report begins with a shutdown review meeting. A report on the turnaround is prepared by the maintenance head. This report is circulated to all heads of departments both as a motivating device and for learning to overcome deficiencies. The report records the following.

Achievements during the turnaround: This can be problems faced and successfully tackled and other good performances such as extraordinary performance of individuals from the points of perseverance, problem solving, conflict resolution, collaborative efforts, productivity increases, etc.

- Good performance in planning by way of good estimates of time for various activities and resources needed as well as excellent monitoring so that delays could be spotted and corrective actions taken in time.
- Good performance of contractors inclusive of their coming forward to tackle various problem issues and accelerating efforts required by the company.
- Good performance of vendors inclusive of their coming to the rescue of the company in terms of quick delivery of materials required in a rush for the turnaround.

Deficiencies noted during the turnaround: These may be in the following areas:

System deficiencies: Deficiencies may be observed based on some failure or incident or some points raised during the review meeting. System deficiencies may also be noted in various areas of the turnaround such as shortlisting of contractors, vendor evaluation, material inspection, and payments to the vendor.

Deficiencies in procedures: Similarly, any deficiency in procedure observed or highlighted during the review meetings should be presented so that the same can be resolved at the appropriate level.

Deficiencies in equipment: Shortfall in equipment required or capacity of equipment used should be mentioned.

Training of personnel: Any incident that came up highlighting the need for training of the personnel deployed should be highlighted so that the same may be considered in the future turnarounds.

Lesson Learned Session

In addition to communicating the closure of a shutdown in writing, it is also advisable to have a mechanism for group review. A "lessons learned" session is a valuable closure mechanism for team members, regardless of the shutdown's success. Some typical questions to answer in such a session include

- Did the delivered product meet the specified requirements and goals of the shutdown?
- Was the customer satisfied with the end product?
- Were cost budgets met?
- Was the schedule met?
- Were risks identified and mitigated?
- Did the shutdown management methodology work?
- What could be done to improve the process?
- What bottlenecks or hurdles were experienced that impacted the shutdown?
- What procedures should be implemented in future shutdowns?
- What can be done in future shutdowns to facilitate success?
- What changes would assist in speeding up future shutdowns while increasing communication?

The lessons learned session is typically a meeting that includes

- Shutdown team
- Stakeholder representation including external shutdown oversight, auditor, or quality assurance (QA)
- Executive management (optional)
- Maintenance and operations staff
- Shutdown sponsor (optional)

Such a session provides official closure to a shutdown. It also provides a forum for team member recognition and offers an opportunity to discuss ways to improve future processes and procedures.

Improvements: The review should also target improvements if required over existing procedures, processes, etc., and this should be done by asking the following questions:

- What contributed most to the success (failure) of this shutdown? What worked well?

- What did not work well? What were the constraints that limited our performance?
- Where did we have problems? Should we have foreseen these? Were there indicators that we missed?
- Did we try to be innovative? Were the innovations effective? How do we know this?
- Are there any other things we should be doing?

Some key questions that may be asked are

- Were the success factors achieved?
- Did the shutdown reach its goals and objectives?
- Was the shutdown on time and budget?
- Were changes handled in a controlled manner?
- Was communication clear, concise, and timely throughout the shutdown?
- Do the stakeholders, particularly customers, view the shutdown in a positive manner?
- Was the shutdown well managed?
- Did the team work well together?
- Did the team pull together when facing struggles and solve the problems?

Celebrating

Celebrating the success of the shutdown is important for the closure of the shutdown. The celebration is a time to recognize the workers for what they have contributed and accomplished.

There is fairly universal recognition that positive reinforcement, or rewarding behavior, is an effective management tool. Since it is a goal within the state to execute all shutdowns successfully, it is important to recognize teams that have met this goal. When success in a shutdown is achieved, be certain to provide some recognition to the team. If individuals are singled out for significant achievements, don't forget to recognize the entire team as well.

Management may also want to express recognition of a successful team effort by praising the team at a key meeting or a large gathering of staff. People are proud to have senior management appreciation stated, and such recognition sets the stage for future successful work.

18

Performance

Performance Measure: Success

The organization must measure turnaround performance and observe trends. As with all measurements, a single indicator can mislead. It is, therefore, necessary to design a number of criteria to provide a balanced indication of performance of the shutdown. Table 18.1 lists some suggested criteria.

Having a work process does not guarantee a successful turnaround, but benchmarking considerably reduces the likelihood of failure. Organizations that complete turnarounds on time, on budget, and without surprise invariably have a defined work process and adhere to it.

Though the process is the key; today's technology provides tools that allow organizations to specify the process, define the tasks, and measure adherence. Web-based technology is an ideal medium to organize and control the multitude of tasks, information, and issues that are critical to the successful completion of turnarounds.

Performance check: The performance of all equipment should be recorded. This serves two purposes: one, whether it is known that the equipment is performing as expected, and the other, for the base readings for all equipment for their performance levels. Performance checks should include

- Visual observations of critical equipment
- Online vibration measurements on equipment and also critical pipelines
- Skin temperature measurements of internally refractory lined equipment
- Record flows, yield points, and other data for reference for comparing post-turnaround results in the future

TABLE 18.1

Performance Criteria

Criterion	Description
Duration	Oil out to on-spec product, days or day/year
Total cost	Turnaround and routine maintenance
Turnaround costs	Actual and annualized by plant function
Frequency	Run length, months
Predictability	Actual versus planned work hours, duration, and cost
Safety	Accident number and rates
Start-up incidents	Days lost due to rework
Unscheduled shutdowns	Days lost during the run
Mechanical availability	Time available, percentage
Additional work	Actual versus contingency
Environmental incidents	Impact of incidents attributable to a shutdown
Savings	Money saved resulting from changes to above indices

Lean Shutdowns: Cut Waste Wisely from Your Outages

For companies that run continuously, shutdowns and outages consume a lion's share of maintenance and capital budgets. By its very nature, the shutdown is "fat." The reason for this is the skewed balance between downtime expenses and the cost of shutdown resources. In some cases, having extra resources, such as cranes, are dwarfed by the avoided cost of downtime. Shutdowns are also fat because of an attitude that includes worrying about the budget when it's over.

Times have changed, however, and it's important to rethink your approach to shutdowns. All parts of the organization now come with budget scrutiny. As a result, we're running our shutdowns under tremendous pressure.

One strong temptation is to cut corners on safety or environmental issues. For example, I can't imagine skimping on fall protection harnesses or safety glasses. Small improvements in managing shutdowns can provide large weight losses for the maintenance department. The key is to cut waste without compromising safety or environmental compliance.

Based on my experience from a variety of shutdowns, here are some important actions to take.

1. *Estimates are to be perfect*: Maintenance shutdowns and turnarounds require maintenance personnel to complete a lot of work in a short period. As a result, maintenance planners may feel pressurized to achieve perfect estimates of labor hours for jobs. However, estimates often cannot account for unforeseen situations that are out of a planner's control.

2. *Reduce surprises*: At the beginning of a shutdown (when you start opening things up), you may discover some surprises. Some ideas to reduce this occurrence include

- Open everything on day one.
- Keep a history based on previous experience. Such a history will be important as it will show any deterioration in efficiency. A lot of things deteriorate at a relatively constant pace and they have similar failure modes.
- Diagnostic predictive maintenance technology (i.e., infrared and vibration analysis) might give an indication of what's going on. Schedule nondestructive testing (NDT) right before you close the work list.
- Do a mini-shutdown before conducting a bigger shutdown. When you do a mini-shutdown, you open everything up, perform inspections, close the job, and go back into service. This isn't possible in a lot of places, but crewmembers said their shutdowns were relatively controlled and didn't produce a lot of surprises.

3. *Use software*: If there are more than 25 tasks in the shutdown (a very small event), then using software for project management will lean up the shutdown by shortening the duration. Be sure planners and schedulers are well trained in the software package you use. The advantages are simple:

- By calculating the critical path, you know early on if the shutdown is on or off schedule.
- By realizing the tasks that are on or near the critical path, you know what to focus on.
- Knowing that without extra intervention, if a critical path item is behind, then the whole shutdown will be late.
- You can see a problem coming when it's small enough to easily fix.
- You can create displays that explain the shutdown and show its current status.

4. *Plan properly*: Did you know that 85% of planning and scheduling is done before the shutdown begins? The point of planning is to identify the elements of a particular, unique job. The main point of scheduling involves precisely bringing together the key elements of a unique specific maintenance job:

- People with the right skills to perform the job and are physically able and mentally alert.
- Safe job steps.
- Correct parts, materials, tools, supplies, and consumables for the job.
- Adequate equipment for lifting, bending, drilling, welding, etc.

- Personal protective equipment (PPE).
- Proper permits and lockouts.
- Custody and control of the asset.
- Safe access to assets, safe work platforms, and humane working conditions.
- Updated drawings and wiring diagrams and other information.
- Proper waste disposal.
- Make sure you take advantage of the time before the job starts to line up all of the elements. Remember if any item is missing, the job will stop or people will improvise, which increases the probability of a problem with quality and safety.

Focus on overall daily work completion: Management should not hold technicians to the planned hours for a single job. Instead, focus on technicians completing all jobs assigned to them for the day.

Good maintenance planners estimate reasonable amount of work for a week and maintenance supervisors can assign reasonable amount of work for each day. Maintenance productivity increases because the plant assigns a reasonable amount of work.

Work assignments and expectations: A maintenance supervisor should assign each maintenance technician a full day's work at the beginning of the day, not one job at a time. Base the amount of work to assign on the maintenance planner's estimates, which should be reasonable. This approach lets the maintenance technicians set their goals of how much work to accomplish each day and helps them pace their workday. This approach reduces the stress of wondering what the next job is and eliminates guessing at expectations.

5. *Don't over order materials*: Keep an eye on the over-ordering of materials and returns as soon as possible. When the shutdown is completed, the tendency is to shove all the extra material into the storeroom and take credit for the value. In this way, the shutdown budget is helped, but there's an overall cost to the organization unless the material is used in a fairly short time. Many storerooms have leftovers from projects and shutdowns for years after the event.

6. *Make certain there are enough supplies*: In terms of whether there are enough supplies for the whole shutdown, the planner should put his or her hands on these items and not accept the computer's inventory level. Supplies include rags, oil-dry compounds, welding rods or wires, gases, nuts and bolts, etc. Shutdowns have been stopped in their tracks because someone made an assumption about simple resources (i.e., running out of welding wire or rod, not having enough torque wrenches).

7. *Limit unneeded rentals*: Keep an eye on the excessive numbers of rented cranes, welding units, generators, compressors, tanks, scaffolding, and other equipment. Investigate and return what's clearly not needed and doesn't provide any benefit, unless it's there to provide insurance against some significant loss. Return equipment rentals of all kinds as soon as practical.

8. *Eliminate unused resources*: Be on the lookout for situations where resources are being paid for, but are not being used because of resource-leveling problems. Have some lean shutdown ready. This would also include spending a little extra to leave scaffolding to do some routine maintenance after the shutdown, or keeping cranes for a few extra days, as well as labor during the shutdown.

9. *Validate the work list*: Validate the work list and remove duplications, remove jobs that aren't essential, and be sure the wording of the work requested is clear. On individual jobs, look at the scope of work as a contractor would. Be sure it's as clear and complete as possible. A better scope will result in lower prices if there are fewer unknowns.

10. *Settle claims with contractors*: Settle claims with your contractors promptly.

11. *Have good, efficient meetings*: Good meetings are essential to the success of the entire shutdown effort, while encouraging the lean process. Wasted meeting time is highly leveraged. If there are eight people at a meeting and they're waiting for the ninth, then the group's time is being wasted. It isn't just one person's time; the loss is leveraged and time involving the eight people is gone. You've got to have productive meetings.

    People come in late for meetings or don't do their homework. They don't pay attention and then act inconsistently with the decisions of the group. Moreover, they don't have good discipline and management often doesn't have a solid behavior model for meetings either. By the way, do you have rules about sending text messages or checking e-mail during meetings?

Appendix A: Pressure Ratings of Steel Pipe

| Pipe | | Pressure—PSI | | | Pipe | | Pressure—PSI | | |
Nom. Size (in.)	Sch. No.	Working	Burst	Water Hammer Factor	Nom. Size (in.)	Sch. No.	Working	Burst	Water Hammer Factor
⅛	40	3500	20,200	—	2½	160	4200	15,700	5.43
⅛	80	4800	28,000	—	2½	XXS	6900	23,000	7.82
¼	40	2100	19,500	—	3	40	1600	7,400	2.60
¼	80	4350	26,400	—	3	80	2600	10,300	2.92
⅜	40	1700	16,200	—	3	160	4100	15,000	3.56
⅜	80	3800	22,500	—	3	XXS	6100	20,500	4.64
½	40	2300	15,600	63.4	3½	40	1500	6,800	1.94
½	80	4100	21,000	—	3½	80	2400	9,500	2.17
½	160	7300	26,700	—	4	40	1400	6,300	1.51
½	XXS	12300	42,100	—	4	80	2300	9,000	1.67
¾	40	2000	12,900	36.1	4	160	4000	14,200	2.08
¾	80	3500	17,600	44.5	4	XXS	5300	18,000	2.47
¾	169	8500	25,000	—	5	40	1300	5,500	0.960
¾	XXS	10000	35,000	—	5	80	2090	8,100	1.06
1	40	2100	12,100	22.3	5	160	3850	13,500	1.32
1	80	3500	15,900	26.8	5	XXS	4780	16,200	1.49
1	160	5700	22,300	36.9	6	40	1210	5,100	0.666
1	XXS	9500	32,700	68.3	6	80	2070	7,800	0.738
1¼	40	1800	10,100	12.9	6	160	3760	13,000	0.912
1¼	80	3000	13,900	15.0	6	XXS	4660	15,000	1.02
1¼	160	4400	18,100	18.2	8	40	1100	4,500	0.385
1¼	XXS	7900	27,700	30.5	8	80	1870	6,900	0.422
1½	40	1700	9,100	9.46	8	160	3700	12,600	0.529
1½	80	2800	12,600	10.9	8	XXS	3560	12,200	0.519
1½	160	4500	17,700	13.7	10	40	1030	4,100	0.244
1½	XXS	7200	25,300	20.3	10	80	1800	6,600	—
2	40	1500	7,800	5.74	10	160	3740	12,500	0.340
2	80	2500	11,000	6.52	10	XXS	3300	11,200	—
2	160	4600	17,500	8.60	12	40	1000	3,800	—
2	XXS	6300	22,100	10.9	12	80	1800	6,500	—

(continued)

(continued)

| Pipe | | Pressure—PSI | | | Pipe | | Pressure—PSI | | |
Nom. Size (in.)	Sch. No.	Working	Burst	Water Hammer Factor	Nom. Size (in.)	Sch. No.	Working	Burst	Water Hammer Factor
2½	40	1900	8,500	4.02	12	160	3700	12,300	0.239
2½	80	2800	11,500	4.54	12	XXS	2700	9,400	—

Note: Based on ASTM A53 Grade B or A106 Grade B, Seamless ANSI B31.1, 1977 with allowances for connections and fittings reduces these working pressures by approximately 25%.

The allowable pressures were calculated by the formula in the Code for Pressure Piping, ASA B31.1–1955, Section 3, par. 324(a):

$$P = \frac{25(t-C)}{D-2y(t-C)}$$

where
   P is the allowable pressure in lb per sq in. (gauge)
   S is the allowable working stress in lb per sq in.
   D is the outside diameter in inches
   t is the design thickness in inches or 12½% less than the nominal thickness shown in the table
   C is the allowance in inches for corrosion and/or mechanical strength (C = 0.05 in. has been used earlier for all pipe sizes)
   y is the a coefficient having values for ferritic steels, as follows:
      0.4 up to and including 900°F
      0.5 for 950°F
      0.7 for 1000°F and above

The allowable working stresses were obtained from the Code for Pressure Piping, ASA B31.1.1–1955, Table 12.

   Hydraulic machinery piping is not covered by the Code for Pressure Piping, but it is current practice to use stresses comparable with those given for Refinery and Oil Transportation Piping, Division A. The allowable working pressures at 100°F tabulated earlier accordingly may be used, provided that water hammer or shock conditions are considered by reducing these values by the product of the flow rate in gallons per minute and the Water Hammer Factor tabulated earlier.

   Thus, if the flow rate is 100 gpm in a 2 in. extra strong line, the shock pressure created by water hammer is 100 × 6.52 = 652 lb/in.[2]; by deducting this from the value of 2500 lb/in.[2] shown in the table, the allowable static working pressure is found to be 1848 lb/in.[2].

Burst pressures for pipe were calculated using formula

$$P = \frac{25t}{OD}$$

where
P is the internal burst pressure, psig
S is the allowable stress (60,000 psi)
OD is the outside diameter of tube in inches
t is the nominal wall thickness

Steel Pipe: Size, Schedule, and Flow Rates

Standard Pipe—Schedule 40

Pipe Size	OD	Wall	ID	Int Area	WT/FT	GPM @ 2 FPS	GPM @ 5 FPS	GPM @ 10 FPS	GPM @ 15 FPS	GPM @ 20 FPS	GPM @ 25 FPS
⅛	0.405	0.068	0.269	0.057	0.245	0.35	0.89	1.8	2.7	3.5	4.4
¼	0.540	0.088	0.364	0.104	0.425	0.65	1.6	3.2	4.9	6.5	8.1
⅜	0.675	0.091	0.493	0.191	0.567	1.2	3.0	6.0	9.0	12.0	15.0
½	0.840	0.109	0.622	0.304	0.852	1.9	4.8	9.5	12.0	19.0	23.8
¾	1.050	0.113	0.824	0.533	1.132	3.3	8.4	16.7	25.1	33.4	41.8
1	1.315	0.133	1.049	0.864	1.679	5.4	13.5	27.0	40.6	54.1	67.7
1¼	1.660	0.140	1.380	1.495	2.273	9.4	23.4	46.8	70.3	93.7	117
1½	1.900	0.145	1.610	2.036	2.718	12.7	31.9	63.7	95.6	127	159
2	2.375	0.154	2.067	3.356	3.653	21.0	52.5	105	157	210	263
2½	2.875	0.203	2.469	4.788	5.793	30.0	75.0	150	225	300	375
3	3.500	0.216	3.068	7.393	7.575	46.3	116	232	347	463	579
3½	4.000	0.226	3.548	9.886	9.109	61.9	155	310	465	619	774
4	4.500	0.237	4.026	12.73	10.79	79.7	199	399	598	797	997
4½	5.000	247	4.506	15.95	12.54	99.9	250	499	749	998	1249
5	5.563	0.258	5.047	20.01	14.62	125	313	627	940	1253	1567
6	6.625	0.280	6.065	28.89	18.97	181	452	904	1357	1810	2262
7	7.625	0.301	7.023	38.74	23.54	243	607	1213	1820	2427	3033
8	8.625	0.322	7.981	50.03	28.55	313	783	1567	2350	3134	3917
10	10.75	0.365	10.02	78.85	40.48	494	1235	2470	3705	4940	6175
12	12.75	0.406	11.94	111.9	53.56	701	1753	3506	5259	7012	8765

Extra Strong Pipe—XS—Schedule 80

Wall	ID	Int Area	WT/ FT	GPM @ 2 FPS	GPM @ 5 FPS	GPM @ 10 FPS	GPM @ 15 FPS	GPM @ 20 FPS	GPM @ 25 FPS	Pipe Size
0.095	0.215	0.036	0.314	0.23	0.57	1.1	1.7	2.3	2.8	⅛
0.119	0.302	0.072	0.535	0.45	1.1	2.2	3.4	4.5	5.6	¼
0.126	0.423	0.141	0.738	0.88	2.2	4.4	6.6	8.8	11.0	⅜
0.147	0.546	0.234	1.087	1.5	3.7	7.3	11.0	14.7	18.3	½
0.154	0.742	0.433	1.473	2.7	6.8	13.6	20.3	27.1	33.9	¾
0.179	0.957	0.719	2.171	4.5	11.3	22.5	33.8	45.0	56.3	1
0.191	1.278	1.283	2.996	8.0	20.0	40.1	60.2	80.3	100	1¼
0.200	1.500	1.767	3.631	11.1	27.7	55.3	83.0	110	138	1½
0.218	1.939	2.953	5.022	18.5	46.2	92.5	139	185	231	2
0.276	2.323	4.238	7.661	26.5	66.4	133	199	265	332	2½
0.300	2.900	6.605	10.25	41.4	103	207	310	414	517	3
0.318	3.364	8.888	12.50	55.7	139	278	418	557	696	3½
0.337	3.826	11.50	14.98	72.0	180	360	540	720	900	4
0.355	4.290	14.45	17.61	90.5	226	453	679	905	1132	4½
0.375	4.813	18.19	20.78	114	285	570	855	1140	1425	5
0.432	5.761	26.07	28.57	163	408	816	1225	1633	2041	6
0.500	6.625	34.47	38.05	216	540	1080	1620	2160	2699	7
0.500	7.625	45.66	43.39	286	715	1430	2145	2861	3576	8
0.594	9.562	71.81	64.40	450	1125	2249	3374	4498	5623	10
0.688	11.37	101.61	88.57	636	1591	3182	4774	6365	7956	12

Schedule 160 Pipe

Pipe Size	OD	Wall	ID	Int Area	WT/FT	GPM @ 2 FPS	GPM @ 5 FPS	GPM @ 10 FPS	GPM @ PS	GPM @ 20 FPS	GPM @ 25 FPS
½	0.840	0.187	0.466	1.71	1.310	1.07	2.67	5.34	8.01	10.7	13.4
¾	1.050	0.218	0.587	0.271	1.940	1.70	4.24	8.49	12.7	17.0	21.2
1 in.	1.315	0.250	0.815	0.522	2.850	3.27	8.17	16.3	24.5	32.7	40.8
1¼	1.660	0.250	1.160	1.060	3.764	6.62	16.6	33.1	49.7	66.2	82.8
1½	1.900	0.281	1.338	1.410	4.862	8.81	22.0	44.0	66.1	88.1	110
2	2.375	0.343	1.689	2.241	7.450	14.0	35.1	70.2	105	140	175
2½	2.875	0.375	2.125	3.542	10.01	22.2	55.5	111	167	222	278
3	3.500	0.437	2.626	5.416	14.30	33.9	84.8	170	254	339	424
4	4.500	0.531	3.438	9.283	22.52	58.2	145	291	436	582	727
5	5.563	0.625	4.313	14.61	33.0	91.5	229	458	686	915	1144
6	6.625	0.718	5.189	21.15	45.30	132	331	662	994	1325	1656
8	8.625	0.906	6.813	36.44	74.70	230	571	1142	1713	2384	2855
10	10.75	1.125	8.500	56.75	115.64	355	889	1777	2666	3555	4443
12	12.75	1.312	10.126	80.53	160.33	504	1261	2523	3784	5045	6306

Double Extra Strong Pipe

Wall	ID	Int Area	WT/FT	GPM @ 2 FPS	GPM @ 5 FPS	GPM @ 10 FPS	GPM @ 15 FPS	GPM @ 20 FPS	GPM @ 25 FPS	Pipe Size
0.294	0.252	0.050	1.714	0.32	0.79	1.6	2.4	3.1	3.9	½
0.308	0.434	0.148	2.440	0.93	2.3	4.6	6.9	9.2	11.6	¾
0.358	0.599	0.282	3.659	1.8	4.4	8.8	13.3	17.7	22.1	1 in.
0.382	0.896	0.630	5.214	4.0	9.9	19.8	29.6	39.5	49.4	1¼
0.400	1.100	0.950	6.408	6.0	14.9	29.8	44.6	59.5	74.4	1½
0.436	1.503	1.774	9.029	11.1	27.9	55.6	83.4	111	139	2
0.552	1.771	2.463	13.70	15.4	38.6	77.1	116	154	193	2½
0.600	2.300	4.154	18.58	26.0	65.1	130	195	260	325	3
0.674	3.152	7.803	27.54	48.9	122	244	367	488	611	4
0.750	4.063	12.97	38.55	81.2	203	406	609	812	1015	5
0.864	4.897	18.83	53.16	118	295	590	885	1180	1475	6
0.875	6.875	37.12	72.42	233	581	1163	1744	2325	2907	8
1.000	8.750	60.13	104.1	377	942	1883	2825	3767	4709	10
1.000	10.75	90.76	125.5	569	1421	2843	4264	5686	7107	12

Appendix B: Gaskets

Ring Joint Gasket

The metal ring-joint gaskets have been designed to withstand exceptionally high assembly loads over a small area, thus producing high seating stresses. Ring-joint gaskets are manufactured from a metallic material. The demands on the geometrical accuracy and the surface quality are therefore high. This concerns both the gasket and the sealing section of the flange. The necessary surface quality depends substantially on the Brinell hardness of the gasket material.

R-type ring

BX-type ring

RX-type ring

Material

Standard materials are soft iron, SS316, 316L, 321, 327, low carbon steel, alloy steel f5 and 410, and stainless steel 304 and 304L.

Nonstandard materials used are high nickel alloy and superalloy steel.

Standard Materials

See Table B.1

TABLE B.1

Standard Materials Recommended by the ANSI B16.20

ASTM	DIN Material No.	Maximum HB	Maximum HV	Material Code
Soft iron	1.1003	90	56	D
Low CS	1.0038	120	68	S
4–6 Cr ½ Mo	1.7362	130	72	F5
AISI 410	1.4000	170	86	S 410
AISI 304	1.4301	160	83	S 304
AISI 316	1.4401	160	83	S 316
AISI 347	1.4550	160	83	S 347

Dimensions

See Table B.2

TABLE B.2

Standards for Ring-Joint Gaskets Used with Flanges: Dimensions

Ring-Joint Gasket Style	Ring-Joint Gasket Standard	Flange Standard
R	ASME B 16.20	ANSI B 16.5
	API 6A	ANSI B 16.47 series A
RX	ASME B 16.20	API 6B
	API 6A	
BX	API 6A	API 6BX

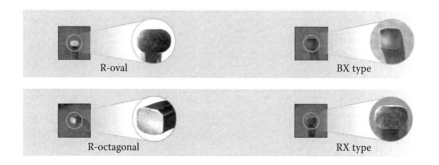

R-oval   BX type

R-octagonal   RX type

Important

- The surface pressure increases or decreases proportional to a possible change of the bolt force.
- Ring-joint gaskets according to API and ASME standards are mainly used in the petrochemical industry and in refineries as a reliable gasket for pipelines with products.

Ring-Joint Gaskets, Type R

Ring-joint gaskets, type R, dimensions according to ASME B16.20 and API STD 6 A for flanges to ASME B16.5 and ASME B16.47, series A (Table B.3).

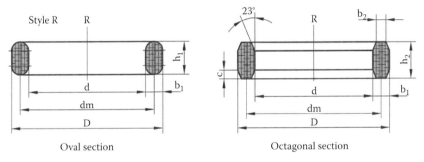

Oval section

Octagonal section

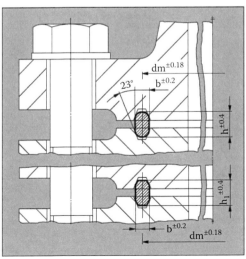

TABLE B.3

Dimension for Ring-Joint Gasket (Type R)

			Dimensions in mm			
			Ring Dimensions			
NPS	Class	Ring No.	dm	b	h	h₁

NPS	Class	Ring No.	dm	b	h	h_1
3	300–600	R 30	117.47	11.11	17.46	15.87
3	300–900	R 31	123.82	11.11	17.46	15.87
3	2500	R 32	127.0	12.7	19.05	17.46
3½	150	R 33	131.76	7.93	14.28	12.7
3½	300–600	R 34	131.76	11.11	17.46	15.87
3	1500	R 35	136.52	11.11	17.46	15.87
4	150	R 36	149.22	7.93	14.28	12.7
4	300–900	R 37	149.22	11.11	17.46	15.87
4	2500	R 38	157.16	15.87	22.22	20.64
4	1500	R 39	161.92	11.11	17.46	15.87
5	150	R 40	171.45	7.93	14.28	12.7
5	300–900	R 41	180.97	11.11	17.46	15.87
5	2500	R 42	190.5	19.05	25.4	23.81
6	150	R 43	193.67	7.93	14.28	12.7
5	1500	R 44	193.67	11.11	17.46	15.87
6	300–900	R 45	211.13	11.11	17.46	15.87
6	1500	R 46	211.13	12.7	19.05	17.46
6	2500	R 47	228.6	19.05	25.4	23.81
8	150	R 48	247.65	7.93	14.28	12.7
8	300–900	R 49	269.87	11.11	17.46	15.87
8	1500	R 50	269.87	15.87	22.22	20.64
8	2500	R 51	279.4	22.22	28.57	26.99
10	150	R 52	304.8	7.93	14.28	12.7
10	300–900	R 53	323.85	11.11	17.46	15.87
10	1500	R 54	323.85	15.87	22.22	20.64
10	2500	R 55	342.9	28.57	36.51	34.92
12	150	R 56	381.0	7.93	14.28	12.7
12	300–900	R 57	381.0	11.11	17.46	15.87
12	1500	R 58	381.0	22.22	28.57	26.99
14	150	R 59	396.87	7.93	14.28	12.7
12	2500	R 60	406.4	31.75	39.68	38.1
14	300–600	R 61	419.1	11.11	17.46	15.87
14	900	R 62	419.1	15.87	22.22	20.64
14	1500	R 63	419.1	25.4	33.33	31.75
16	150	R 64	454.0	7.93	14.28	12.7
16	300–600	R 65	469.9	11.11	17.46	15.87
16	900	R 66	469.9	15.87	22.22	20.64
16	1500	R 67	469.9	28.57	36.51	34.92
18	150	R 68	517.52	7.93	14.28	12.7

TABLE B.3 (continued)

Dimension for Ring-Joint Gasket (Type R)

			Dimensions in mm			
			Ring Dimensions			
NPS	Class	Ring No.	dm	b	h	h_1
18	300–600	R 69	533.4	11.11	17.46	15.87
18	900	R 70	533.4	19.05	25.4	23.81
18	1500	R 71	533.4	28.57	36.51	34.92
20	150	R 72	558.8	7.93	14.28	12.7
20	300–600	R 73	584.2	12.7	19.05	17.46
20	900	R 74	584.2	19.05	25.4	23.81
20	1500	R 75	584.2	31.75	39.68	38.1
24	150	R 76	673.1	7.93	14.28	12.7
24	300–600	R 77	692.15	15.87	22.22	20.64
24	900	R 78	692.15	25.4	33.33	31.75
24	1500	R 79	692.15	34.92	44.45	41.27
22	150	R 80	615.95	7.93		12.7

Spiral Wound Gasket

Properties and Application

Spiral wound gasket (SWG) is a special semimetallic gasket of great resilience; therefore, they are very suitable for application featuring heavy operating condition.

Spiral wounded gasket is manufactured by spirally wounding a v-shaped metal strip and a strip of nonmetallic filler material; the metal strip holds the filler providing the gasket with mechanical resistance and resilience. Spiral wound gasket can be made by an outer centering ring and/or inner retaining ring. The outer centering ring controls the compression and holds the gasket centrally within the bolt circle. The inner retaining ring increases the axial rigidity and resilience of the gasket.

Shape and Construction

Spiral wound gaskets are produced in several styles and combinations of material to fit the most stringent application. Spiral wound gaskets are usually of circular shape; however, one can produce the same in other shapes also, like oval, rectangular, and with round corners.

Gasket Standard Style

- Gasket with guide and inner ring
- Gasket without guide and inner ring
- Gasket with inner ring
- Gasket with guide (outer) ring
- Gasket with guide and with inner ring

Metallic Strip

Standard thickness of the metallic strips is 0.2 mm (0.18) (Table B.4).
  Winding strip materials
  Standard materials used are stainless steel types 304 and 316L.
  Other materials used are stainless steel types 304L, 309, 310, 316Ti, 317L, 321, 347, 430, 17-7PH, alloy 20, Monel, titanium, nickel, copper, and phos bronze.

Filler

Filler is generally used for thickness from 0.5 to 0.6 mm:
- Flexible graphite 98%
- Flexible graphite 99.85%
- Polytetrafluoroethylene (PTFE)
- Ceramic
- Micalit

Centering Ring

The centering ring (CR) does not come direct into contact with contained fluid; it is normally made of carbon steel (CS) and is electroplated and painted to avoid corrosion.

Inner Rings

Inner ring is used to avoid excessive compression due to high seating stress in high-pressure service, and it is also used to reduce turbulence in the flange area; it is normally made of same material as of gasket metallic strip.

TABLE B.4

Materials for Metallic Strip

ASTM	DIN Material No.
AISI 304	1.4301
AISI 316, 316 L	1.4401, 1.4404
AISI 321	1.4541
AISI 316 Ti	1.4571
Monel (NiCu30Fe)	2.4360

Guide Ring Materials

CS is used as standard material.
  Other materials are

- Stainless steel types 304, 304L, 316, 316L, 316Ti, 310, 321, and 347
- Monel
- Titanium
- Nickel

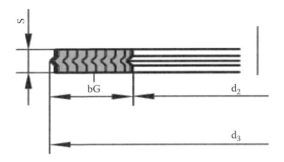

Thickness

The standard manufacturing thickness of spiral wound gasket is 3.2, 4.5, and 6.5 mm (measured across the metallic strip not including the filler, which produces 0.2–0.3 mm beyond the metal).

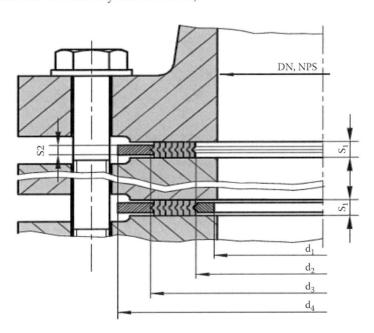

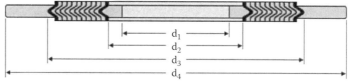

Dimensions of spiral wound gaskets ASME B16.20
used with raised face flanges ASME B16.5

General Notes

Dimensions are in millimeters unless otherwise indicated.

Image shows a spiral wound gasket with inner and outer ring (Table B.5 through B.14):

d_1 is the inside diameter when inner ring is used
d_2 is the inside diameter sealing element when no inner ring is used
d_3 is the outside diameter of sealing element
d_4 is the outside diameter of outer ring

TABLE B.5

Pressure Class 150–NPS 1/2–NPS 24

NPS	Inner Ring Inside (d_1) Diameter	Sealing Element Inside (d_2) Diameter	Sealing Element Outside (d_3) Diameter	Outer Ring Outside (d_4) Diameter
½	14.2	19.1	31.8	47.8
¾	20.6	25.4	39.6	57.2
1	26.9	31.8	47.8	66.8
1¼	38.1	47.8	60.5	76.2
1½	44.5	54.1	69.9	85.9
2	55.6	69.9	85.9	104.9
2½	66.5	82.6	98.6	124
3	81	101.6	120.7	136.7
4	106.4	127	149.4	174.8
5	131.8	155.7	177.8	196.9
6	157.2	182.6	209.6	222.3
8	215.9	233.4	263.7	279.4
10	268.2	287.3	317.5	339.9
12	317.5	339.9	374.7	409.7
14	349.3	371.6	406.4	450.9
16	400.1	422.4	463.6	514.4
18	449.3	474.7	527.1	549.4
20	500.1	525.5	577.9	606.6
24	603.3	628.7	685.8	717.6

TABLE B.6

Pressure Class 300–NPS 1/2–NPS 24

NPS	Inner Ring Inside (d_1) Diameter	Sealing Element Inside (d_2) Diameter	Sealing Element Inside (d_3) Diameter	Outer Ring Inside (d_4) Diameter
½	14.2	19.1	31.8	54.1
¾	20.6	25.4	39.6	66.8
1	26.9	31.8	47.8	73.2
1¼	38.1	47.8	60.5	82.6
1½	44.5	54.1	69.9	95.3
2	55.6	69.9	85.9	111.3
2½	66.5	82.6	98.6	130.3
3	81	101.6	120.7	149.4
4	106.4	127	149.4	181.1
5	131.8	155.7	177.8	215.9
6	157.2	182.6	209.6	251
8	215.9	233.4	263.7	308.1
10	268.2	287.3	317.5	362
12	317.5	339.9	374.7	422.4
14	349.3	371.6	406.4	485.9
16	400.1	422.4	463.6	539.8
18	449.3	474.7	527.1	596.9
20	500.1	525.6	577.9	654.1
24	603.3	628.7	685.8	774.7

TABLE B.7

Pressure Class 400–NPS 1/2–NPS 24

NPS	Inner Ring Inside (d_1) Diameter	Sealing Element Inside (d_2) Diameter	Sealing Element Inside (d_3) Diameter	Outer Ring Inside (d_4) Diameter
½	14.2	19.1	31.8	54.1
¾	20.6	25.4	39.6	66.8
1	26.9	31.8	47.8	73.2
1¼	38.1	47.8	60.5	82.6
1½	44.5	54.1	69.9	95.3
2	55.6	69.9	85.9	111.3
2½	66.5	82.6	98.6	130.3
3	81	101.6	120.7	149.4
4	102.6	120.7	149.4	177.8
5	128.3	147.6	177.8	212.9
6	154.9	174.8	209.6	247.7
8	205.7	225.6	263.7	304.8
10	255.3	274.6	317.5	358.9
12	307.3	327.2	374.7	419.1
14	342.9	362	406.4	482.6
16	389.9	412.8	463.6	536.7
18	438.2	469.9	527.1	593.9
20	489	520.7	577.9	647.7
24	590.6	628.7	685.8	768.4

TABLE B.8

Pressure Class 600–NPS 1/2–NPS 24

NPS	Inner Ring Inside (d_1) Diameter	Sealing Element Inside (d_2) Diameter	Sealing Element Outside (d_3) Diameter	Outer Ring Outside (d_4) Diameter
½	14.2	19.1	31.8	54.1
¾	20.6	25.4	39.6	66.8
1	26.9	31.8	47.8	73.2
1¼	38.1	47.8	60.5	82.6
1½	44.5	54.1	69.9	95.3
2	55.6	69.9	85.9	111.3
2½	66.5	82.6	98.6	130.3
3	78.7	101.6	120.7	149.4
4	102.6	120.7	149.4	193.8
5	128.3	147.6	177.8	241.3
6	154.9	174.8	209.6	266.7
8	205.7	225.6	263.7	320.8
10	255.3	274.6	317.5	400.1
12	307.3	327.2	374.7	457.2
14	342.9	362	406.4	492.3
16	389.9	412.8	463.6	565.2
18	438.2	469.9	527.1	612.9
20	489	520.7	577.9	682.8
24	590.6	628.7	685.8	790.7

TABLE B.9

Pressure Class 900–NPS 1/2–NPS 24

NPS	Inner Ring Inside (d_1) Diameter	Sealing Element Inside (d_2) Diameter	Sealing Element Outside (d_3) Diameter	Outer Ring Outside (d_4) Diameter
½	14.2	19.1	31.8	63.5
¾	20.6	25.4	39.6	69.9
1	26.9	31.8	47.8	79.5
1¼	33.3	39.6	60.5	88.9
1½	41.4	47.8	69.9	98.6
2	52.3	58.7	85.9	143
2½	63.5	69.9	98.6	165.1
3	78.7	95.3	120.7	168.4
4	102.6	120.7	149.4	206.5
5	128.3	147.6	177.8	247.7
6	154.9	174.8	209.6	289.1
8	196.9	222.3	257.3	358.9
10	246.1	276.4	311.2	435.1
12	292.1	323.9	368.3	498.6
14	320.8	355.6	400.1	520.7
16	374.7	412.8	457.2	574.8
18	425.5	463.6	520.7	638.3
20	482.6	520.7	571.5	698.5
24	590.6	628.7	679.5	838.2

TABLE B.10

Pressure Class 1500–NPS 1/2–NPS 24

NPS	Inner Ring Inside (d_1) Diameter	Sealing Element Inside (d_2) Diameter	Outside (d_3) Diameter	Outer Ring Outside (d_4) Diameter
½	14.2	19.1	31.8	63.5
¾	20.6	25.4	39.6	69.9
1	26.9	31.8	47.8	79.5
1¼	33.3	39.6	60.5	88.9
1½	41.4	47.8	69.9	98.6
2	52.3	58.7	85.9	143
2½	63.5	69.9	98.6	165.1
3	78.7	92.2	120.7	174.8
4	97.8	117.6	149.4	209.6
5	124.5	143	177.8	254
6	147.3	171.5	209.6	282.7
8	196.9	215.9	257.3	352.6
10	246.1	266.7	311.2	435.1
12	292.1	323.9	368.3	520.7
14	320.8	362	400.1	577.9
16	374.7	406.4	457.2	641.4
18	425.5	463.6	520.7	704.9
20	476.3	514.4	571.5	755.7
24	577.9	616	679.5	901.7

TABLE B.11

Pressure Class 2500–NPS 1/2–NPS 12

NPS	Inner Ring Inside (d_1) Diameter	Sealing Element Inside (d_2) Diameter	Outside (d_3) Diameter	Outer Ring Outside (d_4) Diameter
½	14.2	19.1	31.8	69.9
¾	20.6	25.4	39.6	76.2
1	26.9	31.8	47.8	85.9
1¼	33.3	39.6	60.5	104.9
1½	41.4	47.8	69.9	117.6
2	52.3	58.7	85.9	146
2½	63.5	69.9	98.6	168.4
3	78.7	92.2	120.7	196.9
4	97.8	117.6	149.4	235
5	124.5	143	177.8	279.4
6	147.3	171.5	209.6	317.5
8	196.9	215.9	257.3	387.4
10	246.1	270	311.2	476.3
12	292.1	317.5	368.3	549.4

TABLE B.12

SWG for BS 1560 and ASME B 16.5 Flanges

NPS (in.)	d₁ (mm)		d₂ (mm)		d₃ (mm)	d₄ (mm)						
Class (lb)	150–400	600–2500	150–400	600–2500	150–2500	150	300	400	600	900	1500	2500
½	12.7	12.7	19.1	19.1	31.8	44.4	50.8	50.8	50.8	60.3	60.3	66.7
¾	20.6	20.6	27	27	39.7	53.9	63.5	63.5	63.5	66.7	66.7	73
1	27	27	33.3	33.3	47.6	63.5	69.8	69.8	69.8	76.2	76.2	82.5
1¼	41.3	39.7	47.6	46	60.3	73	79.4	79.4	79.4	85.7	85.7	101.6
1½	49.2	47.6	55.6	54	69.9	82.5	92.1	92.1	92.1	95.2	95.2	114.3
2	61.9	60.3	71.4	69.9	85.7	101.6	108	108	108	139.7	139.7	142.8
2½	74.6	73	84.1	82.6	98.4	120.6	127	127	127	161.9	161.9	165.1
3	95.3	92.1	104.8	101.6	120.7	133.4	146.1	146.1	146.1	165.1	171.5	193.7
3½	108	104.8	117.5	114.3	133.4	158.8	161.9	158.7	158.7	—	—	—
4	117.5	114.3	130.2	127	149.2	171.5	177.8	174.6	190.5	203.2	206.4	231.7
5	144.5	141.3	157.2	154	177.8	193.7	212.7	209.5	238.1	244.5	250.8	276.2
6	171.5	168.3	184.2	181	209.6	219.1	247.7	244.5	263.5	285.8	279.4	314.3
8	222.3	219.1	235	231.8	263.5	276.2	304.8	301.6	317.5	355.6	349.3	384.1
10	276.2	269.9	288.9	282.6	317.5	336.5	358.8	355.6	396.9	431.8	431.8	473
12	330.2	323.8	342.9	336.5	374.6	406.4	419.1	415.9	454	495.3	517.5	546.1
14	361.9	355.6	374.6	368.3	406.4	447.7	482.6	479.4	488.9	517.5	574.7	—
16	412.7	406.4	425.4	419.1	463.5	511.2	536.6	533.4	561.9	571.5	638.1	—
18	466.7	460.4	479.4	473.1	527	546.1	593.7	590.5	609.6	635	701.7	—
20	517.5	511.2	530.2	523.9	577.8	603.2	650.9	644.5	679.5	695.3	752.4	—
22	574.4	568.4	587.4	581.1	635	657.2	701.7	698.5	730.3	—	—	—
24	622.3	615.9	635	628.6	685.8	714.4	771.5	765.2	787.4	835	898.5	—

TABLE B.13

ASME B 16.20 Gaskets for ASME B 16.5 Flanges

NPS (in.) Class (lb)	d_1 (mm) 150–300	400–600	900	1500	2500	d_2 (mm) 150–300	400–600	900	1500	2500	d_3 (mm) 150–600	900–2500	d_4 (mm) 150	300	400	600	900	1500	2500
½	14.2	14.2	14.2	14.2	14.2	19.1	19.1	19.1	19.1	19.1	31.8	31.8	47.8	54.1	54.1	54.1	63.5	63.5	69.9
¾	20.6	20.6	20.6	20.6	20.6	25.4	25.4	25.4	25.4	25.4	39.6	39.6	57.2	66.8	66.8	66.8	69.9	69.9	76.2
1	26.9	26.9	26.9	26.9	26.9	31.8	31.8	31.8	31.8	31.8	47.8	47.8	66.8	73.2	73.2	73.2	79.5	79.5	85.9
1¼	38.1	38.1	38.1	33.3	33.3	47.8	47.8	39.6	39.6	39.6	60.5	60.5	76.2	82.6	82.6	82.6	88.9	88.9	104.9
1½	44.5	44.5	44.5	41.4	41.4	54.1	54.1	47.8	47.8	47.8	69.9	69.9	85.9	95.3	95.3	95.3	98.6	98.6	117.6
2	55.6	55.6	55.6	52.3	52.3	69.9	69.9	58.7	58.7	58.7	85.9	85.9	104.9	111.3	111.3	111.3	143.0	143.0	146.1
2½	66.5	66.5	66.5	63.5	63.5	82.6	82.6	69.9	69.9	69.9	98.6	98.6	124.0	130.3	130.3	130.3	165.1	165.1	168.4
3	81.0	81.0	81.0	81.0	81.0	101.6	101.6	95.3	92.2	92.2	120.7	120.7	136.7	149.4	149.4	149.4	168.4	174.8	196.9
4	106.4	106.4	106.4	106.4	106.4	127.0	120.7	120.7	117.6	117.6	149.4	149.4	174.8	181.1	177.8	193.8	206.5	209.6	235.0
5	131.8	131.8	131.8	131.8	131.8	155.7	147.6	147.6	143.0	143.0	177.8	177.8	196.9	215.9	212.9	241.3	247.7	254.0	279.4
6	157.2	157.2	157.2	157.2	157.2	182.6	174.8	174.8	171.5	171.5	209.6	209.6	222.3	251.0	247.7	266.7	289.1	282.7	317.5
8	215.9	209.6	196.9	196.9	196.9	233.4	225.6	222.3	215.9	215.9	263.7	257.3	279.4	308.1	304.8	320.8	358.9	352.6	387.4
10	268.2	260.4	246.1	246.1	246.1	287.3	274.6	276.4	266.7	270.0	317.5	311.2	339.9	362.0	358.9	400.1	435.1	435.1	476.3
12	317.5	317.5	292.1	292.1	292.1	339.9	327.2	323.9	323.9	317.5	374.7	368.3	409.7	422.4	419.1	457.2	498.6	520.7	549.4
14	349.3	349.3	320.8	320.8	—	371.6	362.0	356.6	362.0	—	406.4	400.1	450.9	485.9	482.6	492.3	520.7	577.9	—
16	400.1	400.1	374.7	368.3	—	422.4	412.8	412.8	406.7	—	463.6	457.2	514.4	539.8	536.7	565.2	574.8	641.4	—
18	449.3	449.3	425.5	425.5	—	474.7	469.9	463.6	463.6	—	527.1	520.7	549.4	596.9	593.9	612.9	638.3	704.9	—
20	500.1	500.1	482.6	476.3	—	525.5	520.7	520.7	514.4	—	577.9	571.5	606.6	654.1	647.7	682.8	698.5	755.7	—
24	603.3	603.3	590.6	577.9	—	628.7	628.7	628.7	616.0	—	685.8	679.5	717.6	774.7	768.4	790.7	838.2	901.7	—

TABLE B.14

ASME B 16.20 Gasket for ASME B 16.47 Series B Flanges

NPS (in.)	d₁ (mm)					d₂ (mm)					d₃ (mm)					d₄ (mm)				
Class (lb)	150	300	400	600	900	150	300	400	600	900	150	300	400	600	900	150	300	400	600	900
26	654.1	654.1	654.1	644.7	666.8	673.1	673.1	668.8	663.7	692.2	698.5	711.2	698.5	714.3	749.3	725.4	771.7	746.3	765.3	838.2
28	704.9	704.9	701.8	692.2	717.6	723.9	723.9	714.5	704.9	743.0	749.3	762.0	749.3	755.7	800.1	776.2	825.5	800.1	819.2	901.7
30	755.7	755.7	752.6	752.6	781.1	774.7	774.7	765.3	778.0	806.5	800.1	812.8	806.5	828.8	857.3	827.0	886.0	857.3	879.6	958.3
32	806.5	806.5	800.1	793.8	838.2	825.5	825.5	812.8	831.9	863.6	850.9	863.6	858.5	882.7	914.4	881.1	939.8	911.4	933.5	1016.0
34	857.3	857.3	850.9	850.9	895.4	876.3	876.3	866.9	889.0	920.8	908.1	914.4	911.4	939.8	971.6	935.0	993.9	962.2	997.0	1073.2
36	908.1	908.1	898.7	901.7	920.8	927.1	927.1	917.7	939.8	946.2	958.9	962.2	965.2	990.6	997.0	987.6	1047.8	1022.4	1047.8	1124.0
38	958.9	971.6	952.5	952.5	1009.7	974.6	1009.7	971.6	990.6	1035.1	1009.7	1047.8	1022.4	1041.4	1085.9	1044.7	1098.6	1073.2	1104.9	1200.2
40	1009.7	1003.3	1000.3	1009.7	1060.5	1022.4	1060.5	1025.7	1047.8	1098.6	1063.8	1098.6	1076.5	1098.6	1149.4	1095.5	1149.4	1127.3	1155.7	1251.0
42	1060.5	1054.1	1051.1	1066.8	1111.3	1079.5	1111.3	1076.5	1104.9	1149.4	1114.6	1149.4	1127.3	1155.7	1200.2	1146.3	1200.2	1178.1	1219.2	1301.8
44	1111.3	1124.0	1104.9	1111.3	1155.7	1124.0	1162.1	1130.3	1162.1	1206.5	1165.4	1200.2	1181.1	1212.9	1257.3	1197.1	1251.0	1231.9	1270.0	1368.6
46	1162.1	1178.1	1168.4	1162.1	1219.2	1181.1	1216.2	1193.8	1212.9	1270.0	1224.0	1254.3	1244.6	1263.7	1320.8	1255.7	1317.8	1289.1	1327.2	1435.1
48	1212.9	1231.9	1206.5	1219.2	1270.0	1231.9	1263.7	1244.6	1270.0	1320.8	1270.0	1311.4	1295.4	1320.8	1371.6	1306.6	1368.6	1346.2	1390.7	1485.9
50	1263.7	1267.0	1257.3	1270.0	—	1282.7	1317.8	1295.4	1320.8	—	1325.6	1335.9	1346.2	1371.6	—	1357.4	1419.4	1403.4	1447.8	—
52	1314.5	1317.8	1308.1	1320.8	—	1335.5	1368.6	1346.2	1371.6	—	1376.4	1406.7	1397.0	1422.4	—	1408.2	1470.2	1454.2	1498.6	—
54	1365.3	1365.3	1352.6	1378.0	—	1384.3	1403.4	1403.4	1428.8	—	1422.4	1454.2	1454.2	1479.6	—	1463.8	1530.4	1517.7	1555.8	—
56	1422.4	1428.8	1403.4	1428.8	—	1444.8	1479.6	1454.2	1479.6	—	1477.8	1524.0	1505.0	1503.4	—	1514.6	1593.8	1568.5	1612.9	—
58	1478.0	1484.4	1454.2	1473.2	—	1500.4	1535.2	1505.0	1536.7	—	1528.8	1573.3	1555.8	1587.5	—	1579.6	1655.8	1619.3	1663.7	—
60	1535.2	1557.3	1517.7	1530.4	—	1557.3	1589.0	1568.5	1593.9	—	1586.0	1630.4	1619.3	1644.7	—	1630.4	1706.6	1682.8	1733.6	—

Gasket Compression

Spiral wound gasket is designed in such a way that a uniform bolt stress based on nominal root diameter compresses the gasket to the thickness (e) (Table B.15).

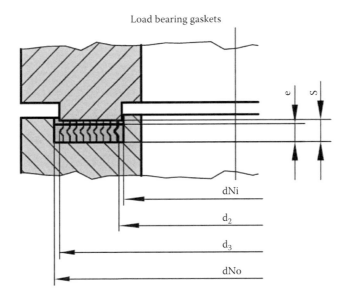

Load bearing gaskets

TABLE B.15

Standard Gasket Compression

S	3.2	4.5	6.5
e	2.5 ± 0.1	3.3 ± 0.1	4.7 ± 0.1

Appendix C: Scaffolding

Scaffolding can provide an efficient and safe means to perform work in the shutdown. However, unsafe scaffolding procedures can lead to accidents, serious injuries, and death. This guide makes clear that planning ahead for the erection, use, and dismantling of scaffolding can substantially reduce scaffold-related accidents and injuries. Compliance with all scaffolding standards will help ensure a safer workplace for employees. Safety and health in the workplace are everyone's responsibility.

Policy for Safe Scaffold Erection and Use

- Sound design
- Selecting the right scaffold for the job
- Assigning personnel
- Training
- Fall protection
- Guidelines for proper erection
- Guidelines for use
- Guidelines for alteration and dismantling
- Inspections
- Maintenance and storage

Sound Design

The scaffold should be capable of supporting its own weight at least four times the maximum intended load to be applied or transmitted to the scaffold and components. Guardrails should be able to withstand at least 200 lb of force on the top rail and 100 lb on the midrail. On complex systems, the services of an engineer may be needed to determine the loads at particular points.

Selecting the Right Scaffold for the Job

You cannot contract away the responsibility for selecting the right scaffold for your job. But if you do contract for scaffolding,

- Choose a scaffold supplier, rental agency, and/or erector who is thoroughly knowledgeable about the equipment needed and its safe use.
- Obtain the owner's manual prepared by the scaffolding manufacturer, which states equipment limitations, special warnings, and intended use and maintenance requirements.

Account for any special features of the building structure in relationship to the scaffold, including distinctive site conditions.

Factor these considerations into your policy:

- Experience of erection and working personnel
- Length and kind of work tasks to be performed
- Weight of loads to be supported
- Hazards to people working on and near the scaffolding
- Needed fall protection
- Material hoists
- Rescue equipment (particularly for suspended scaffolds)
- Weather and environmental conditions
- Availability of scaffolding, components, etc.

Assigning Personnel

Assign a competent person to oversee the scaffold selection, erection, use, movement, alteration, dismantling, maintenance, and inspection. Only assign trained and experienced personnel to work on scaffolding. Be certain they are knowledgeable about the type of scaffolding to be used and about the proper selection, care, and use of fall protection equipment (perimeter protection, fall protection/work positioning belts and full harnesses, lanyards, lifelines, rope grabs, shock absorbers, etc.).

Training

Employees should receive instruction on the particular types of scaffolds that they are to use. Training should focus on proper erection, handling, use, inspection, removal, and care of the scaffolds. Training must also include the installation of fall protection, particularly guardrails, and the proper selection, use, and care of fall arrest equipment.

Site management personnel should also be familiar with correct scaffolding procedures so they can better determine needs and identify deficiencies.

Fall Protection

Guardrails must be installed on all scaffold platforms in accordance with required standards and at least consist of top rails, midrails, and toeboards (if more than 10 ft above the ground or floor). The top edge height of top rails or equivalent member on supported scaffolds manufactured or placed in service after January 1, 2000, shall be installed between 38 and 45 in. above the platform surface. The top edge height on supported scaffolds manufactured and placed in service before January 1, 2000, and on all suspended scaffolds where both a guardrail and a personal fall arrest system are required shall be between 36 and 45 in. When it is necessary to remove guardrails (e.g., to off-load materials), supervision must ensure that they are replaced quickly.

The full-body harness is a belt system designed to distribute the impact energy of a fall over the shoulders, thighs, and buttocks. A properly designed harness will permit prolonged worker suspension after a fall without restricting blood flow, which may cause internal injuries. Rescue is also aided because of the upright positioning of the worker.

General Guidelines for Proper Erection

Accidents and injuries can be reduced when the guidelines in this section are followed.

Supervise the erection of scaffolding. This must be done by a person competent by skill, experience, and training to ensure safe installation according to the manufacturer's specifications and other requirements.

Know the voltage of energized power lines. Ensure increased awareness of location of energized power lines; maintain safe clearance between scaffolds and power lines (i.e., minimum distance of 3 ft for insulated lines less than 300 V; 10 ft for insulated lines 300 V or more). Identify heat sources like steam pipes. Anticipate the presence of hazards before erecting scaffolds and keep a safe distance from them.

Be sure that fall protection equipment is available before beginning erection and use it as needed. Have scaffolding material delivered as close to the erection site as possible to minimize the need for manual handling. Arrange components in the order of erection.

Ensure the availability of material hoisting and rigging equipment to lift components to the erection point and eliminate the need to climb with components. Examine all scaffold components prior to erection. Return and tag "Do Not Use" or destroy defective components.

Prohibit or restrict the intermixing of manufactured scaffold components, unless

1. The components fit together properly, without force.
2. The use of dissimilar metals will not reduce strength.
3. The design load capacities are maintained.

Guidelines for Use

- Be certain that scaffolds and components are not loaded beyond their rated and maximum capacities.
- Prohibit the movement of scaffolds when employees are on them.
- Maintain a safe distance from energized power lines.
- Prohibit work on scaffolds until snow, ice, and other materials that could cause slipping and falls are removed.
- Protect suspension ropes from contact with sources of heat (welding, cutting, etc.) and from acids and other corrosive substances.
- Prohibit scaffold use during storms and high winds.
- Remove debris and unnecessary materials from scaffold platforms.
- Prohibit the use of ladders and other devices to increase working heights on platforms.

Guidelines for Alteration and Dismantling

- Require that scaffolds be altered, moved, and dismantled under the supervision of a competent person.
- Alteration and dismantling activities should be planned and performed with the same care as with erection.
- Tag any incomplete scaffold or damaged component out of service.

Inspections

Inspect all scaffolds and components upon receipt at the erection location. Return, tag "Do Not Use," or destroy defective components. Inspect scaffolds before use and attach a tag stating the time and date of inspection.

Inspect scaffolds before each work shift and especially after changing weather conditions and prolonged interruptions of work. Check for such items as solid foundations, stable conditions, complete working and rest platforms, suitable anchorage points, required guardrails, loose connections, tie-off points, damaged components, proper access, and the use of fall protection equipment.

Maintenance and Storage

Maintain scaffolds in good repair. Only replacement components from the original manufacturer should be used.

Intermixing scaffold components from different manufacturers should be avoided. Fabricated scaffolds should be repaired according to the manufacturer's specifications and guidance. Job-built scaffolds should not be repaired without the supervision of a competent person.

Store all scaffolding parts in an organized manner in a dry and protected environment. Examine all parts and clean, repair, or dispose of them as necessary.

Self-Supporting Scaffolds

A self-supporting scaffold is one or more work platforms supported from below by outriggers, brackets, poles, legs, uprights, posts, frames, or similar supports.

Tube and Coupler

A tube and coupler scaffold is a supported scaffold consisting of platforms supported by individual pieces of tubing, erected with coupling devices

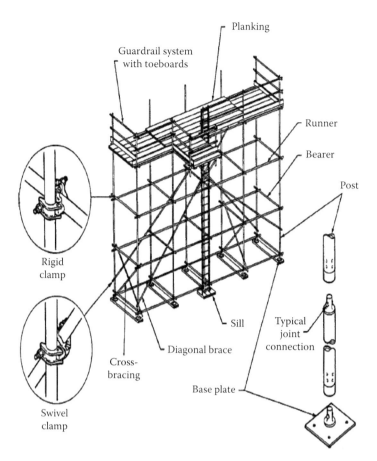

FIGURE C.1
Tube and coupler scaffold.

connecting uprights, braces, bearers, and runners. A registered professional engineer may need to be consulted about the design, construction, and loading of the scaffold. Tube and coupler scaffolds over 125 ft high must be designed by a registered professional engineer and be constructed and loaded consistent with the design (Figure C.1).

Appendix D: Crane and Its Load Chart

Each crane has a load chart that, in short, specifies the crane's capabilities—detailing its features and how its lift capacity varies when considering distance and angle. Just like the old saying "if you fail to plan, you plan to fail," failing to consult a crane load chart before renting or employing a crane for a specific job could leave you with too much or too little capacity for your job.

Before a crane is rented, transported, employed, or purchased, the crane chart must be consulted. Everyone from the crane operator to the job supervisors and even to the sales guys have to know how to read a crane chart.

The important points to be considered in a crane are

Lift capacity—this is where the magic happens. In the legend at the top of the chart, you can see these ratings apply when using 6.5 tons of counterweight, with the outriggers extended to 22 × 22.3 ft. Here, you'd graph out the specific lift the crane is needed for. The "ft." indicator on the left axis represents the radius, the distance from the center pin to the center of the load.

Example: You need to lift a load of 15 tons (30,000 lb) a distance of 25 ft. The distance is measured from the center pin of the crane to the center of the load. Once you determine the distance, look on that line for the largest capacity; that will indicate how many feet of boom must be extended. In this case, it is 45 ft.

It's important to note that the maximum capacity is always measured by the shortest lift, usually over the rear of the crane, and with the outriggers fully extended. While the Terex RT345 has a maximum capacity of 45 tons, lifts at any distance or height drop the maximum capacity dramatically.

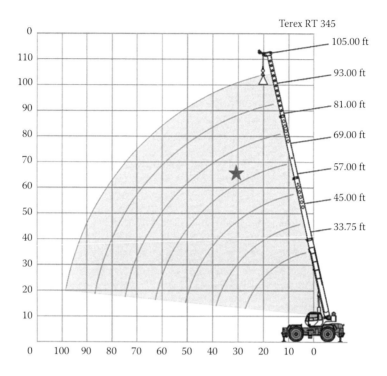

Lift range—just as important as lift capacity is lift range. For that, a range diagram is usually included in every chart, which illustrates how much boom length is needed to pick up and lift a load both at a distance and at height.

Example: You need to pick up a load at 25 ft and lift it to the top of a five story, 65 feet building. Consulting the range diagram, 69 ft of boom is required to make the lift.

Lift angle—With higher angles of lift, the maximum load capacity decreases. With a luffing jib, the angle can be automatically adjusted from the operator cab. With a fixed jib, of course, the angle is fixed.

Appendix E: Shell and Tube Heat Exchangers

Why a Shell and Tube Heat Exchanger?

Shell and tube heat exchangers in their various construction modifications are probably the most widespread and commonly used basic heat exchanger configuration in the process industries. The reasons for this general acceptance are several. The shell and tube heat exchanger provides a comparatively large ratio of heat transfer area to volume and weight. It provides this surface in a form that is relatively easy to construct in a wide range of sizes and that is mechanically rugged enough to withstand normal shop fabrication stresses, shipping and field erection stresses, and normal operating conditions. There are many modifications of the basic configuration, which can be used to solve special problems. The shell and tube exchanger can be reasonably easily cleaned, and those components most subject to failure—gaskets and tubes—can be easily replaced. Finally, good design methods exist, and the expertise and shop facilities for the successful design and construction of shell and tube exchangers are available throughout the world.

Basic Components of Shell and Tube Heat Exchangers

While there is an enormous variety of specific design features that can be used in shell and tube exchangers, the number of basic components is relatively small. These are shown and identified in Figure E.1:

1. *Tubes*. The tubes are the basic component of the shell and tube exchanger, providing the heat transfer surface between one fluid flowing inside the tube and the other fluid flowing across the outside of the tubes. The tubes may be seamless or welded and most commonly made of copper or steel alloys. Other alloys of nickel, titanium, or aluminum may also be required for specific applications.

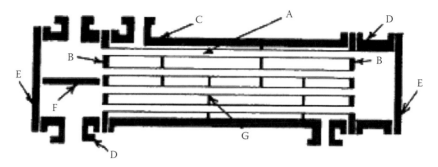

FIGURE E.1
Diagram of typical (fixed tube sheet) shell and tube heat exchanger, showing the components.
A. Tubes; B. Tube sheets; C. Shell and shell-side nozzles; D. Tube-side channels and nozzles;
E. Channel covers; F. Pass divider; G. Baffles.

Tube sheets. The tubes are held in place by being inserted into holes in
the tube sheet and are either expanded into grooves cut into the holes
or welded to the tube sheet where the tube protrudes from the surface.
The tube sheet is usually a single round plate of metal that has been
suitably drilled and grooved to take the tubes (in the desired pattern),
the gaskets, the spacer rods, and the bolt circle where it is fastened to
the shell. However, where mixing between the two fluids (in the event
of leaks where the tube is sealed into the tube sheet) must be avoided.

The tube sheet, in addition to its mechanical requirements, must
withstand corrosive attack by both fluids in the heat exchanger and
must be electrochemically compatible with the tube and all tube-
side material. Tube sheets are sometimes made from low-carbon
steel with a thin layer of corrosion-resisting alloy metallurgically
bonded to one side.

2. *Shell and shell-side nozzles.* The shell is simply the container for the
shell-side fluid, and the nozzles are the inlet and exit ports. The shell
normally has a circular cross section and is commonly made by rolling a metal plate of the appropriate dimensions into a cylinder and
welding the longitudinal joint ("rolled shells"). Small-diameter shells
(up to around 24 in. in diameter) can be made by cutting pipe of the
desired diameter to the correct length ("pipe shells"). The roundness
of the shell is important in fixing the maximum diameter of the baffles
that can be inserted and therefore the effect of shell-to-baffle leakage.
Pipe shells are more nearly round than rolled shells unless particular
care is taken in rolling. In order to minimize out-of-roundness, small
shells are occasionally expanded over a mandrel; in extreme cases,
the shell is cast and then bored out on a boring mill.

In large exchangers, the shell is made out of low-carbon steel
wherever possible for reasons of economy, though other alloys

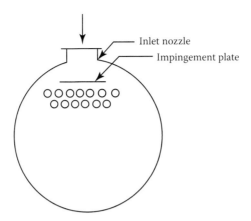

FIGURE E.2
Schematic diagram of placement of an impingement plate under the entrance nozzle.

can be and are used when corrosion or high-temperature strength demands must be met.

The inlet nozzle often has an impingement plate (Figure E.2) set just below to divert the incoming fluid jet from impacting directly at high velocity on the top row of tubes.

Such impact can cause erosion, cavitation, and/or vibration. In order to put the impingement plate in and still leave enough flow area between the shell and plate for the flow to discharge without excessive pressure loss, it may be necessary to omit some tubes from the full circle pattern. Other more complex arrangements to distribute the entering flow, such as a slotted distributor plate and an enlarged annular distributor section, are occasionally employed.

3. *Tube-side channels and nozzles.* Tube-side channels and nozzles simply control the flow of the tube-side fluid into and out of the tubes of the exchanger. Since the tube-side fluid is generally the more corrosive, these channels and nozzles will often be made out of alloy materials (compatible with the tubes and tube sheets, of course). They may be clad instead of solid alloy.

4. *Channel covers.* The channel covers are round plates that bolt to the channel flanges and can be removed for tube inspection without disturbing the tube-side piping. In smaller heat exchangers, bonnets with flanged nozzles or threaded connections for the tube-side piping are often used instead of channels and channel covers.

5. *Pass divider.* A pass divider is needed in one channel or bonnet for an exchanger having two tube-side passes, and they are needed in both channels and bonnets for an exchanger having more than

two passes. If the channels or bonnets are cast, the dividers are integrally cast and then faced to give a smooth bearing surface on the gasket between the divider and the tube sheet. If the channels are rolled from plate or built up from pipe, the dividers are welded in place.

The arrangement of the dividers in multiple-pass exchangers is somewhat arbitrary, the usual intent being to provide nearly the same number of tubes in each pass, to minimize the number of tubes lost from the tube count, to minimize the pressure difference across any one pass divider (to minimize leakage), to provide adequate bearing surface for the gasket, and to minimize fabrication complexity and cost.

6. *Baffles.* Baffles serve two functions: Most importantly, they support the tubes in the proper position during assembly and operation and prevent vibration of the tubes caused by flow-induced eddies, and secondly, they guide the shell-side flow back and forth across the tube field, increasing the velocity and the heat transfer coefficient.

The most common baffle shape is the single segmental, shown in Figure E.3. The segment sheared off must be less than half of the diameter in order to ensure that adjacent baffles overlap at least one full tube row. For liquid flows on the shell side, a baffle cut of 20%–25% of the diameter is common; for low-pressure gas flows, 40%–45% (i.e., close to the maximum allowable cut) is more common, in order to minimize pressure drop.

The baffle spacing should be correspondingly chosen to make the free flow areas through the "window" (the area between the baffle edge and shell) and across the tube bank roughly equal.

For many high-velocity gas flows, the single segmental baffle configuration results in an undesirably high shell-side pressure drop. One way to retain the structural advantages of the segmental baffle and reduce the pressure drop is to use the double segmental baffle.

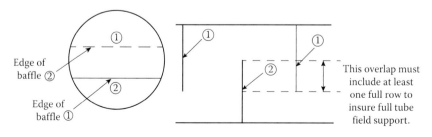

FIGURE E.3
Sketch of typical segmental baffle arrangements.

Provisions for Thermal Stress

1. *Thermal stress problem.* Since, by its very purpose, the shell of the heat exchanger will be at a significantly different temperature than tubes, the shell will expand or contract relative to the tubes, resulting in stresses existing in both components and being transmitted through the tube sheets. The consequences of the thermal stress will vary with circumstances, but shells have been buckled or tubes pulled out of the tube sheet or simply pulled apart. The fixed-tube sheet exchanger shown in Figure E.1 is especially vulnerable to this kind of damage because there is no provision made for accommodating differential expansion.

   There is a rough rule of thumb that says a simple fixed-tube sheet configuration can only be used for cases where the inlet temperatures of the two streams do not differ by more than 100°F. Obviously, there must be many qualifications made to such a flat statement, recognizing the differences in materials and their properties, temperature level of operation, start-up and cycling operational procedures, etc.

2. *Expansion joint on the shell.* The most obvious solution to the thermal expansion problem is to put an expansion roll or joint in the shell, as shown in Figure E.4.

   Internal bellows. In recent years, an internal bellow design has become popular for such applications as waste heat vertical thermosyphon reboilers, where only one pass is permitted on the tube side. These bellows have been designed to operate successfully with high-pressure boiling water on the tube-side and high-temperature reactor effluent gas on the shell.

3. *Floating-Head designs.* Several different designs of "floating-head" shell and tube exchangers are in common use. The goal in each case, of course, is to solve the thermal stress problem and each design does accomplish that goal. Inevitably, however, something must be given up, and each configuration has a somewhat different set of drawbacks to be considered when choosing one.

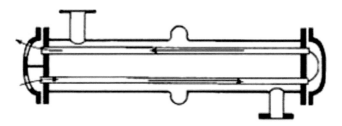

FIGURE E.4
Sketch of fixed tube sheet design incorporating an expansion joint.

The simplest floating-head design is the "pull-through bundle" type. One of the tube sheets is made small enough that it and its gasketed bonnet may be pulled completely through the shell for shell-side inspection and cleaning. The tube side may be cleaned and individual tubes may be replaced without removing the bundle from the shell.

Classification Based on Construction

Fixed-tube sheet: A fixed-tube sheet heat exchanger has straight tubes that are secured at both ends to tube sheets welded to the shell. The construction may have removable channel covers (e.g., AEL), bonnet-type channel covers (e.g., BEM), or integral tube sheets (e.g., NEN) (refer TEMA).

The principal advantage of the fixed-tube sheet construction is its low cost because of its simple construction. In fact, the fixed-tube sheet is the least expensive construction type, as long as no expansion joint is required.

Other advantages are that the tubes can be cleaned mechanically after removal of the channel cover or bonnet and that leakage of the shell-side fluid is minimized since there are no flanged joints.

A disadvantage of this design is that since the bundle is fixed to the shell and cannot be removed, the outsides of the tubes cannot be cleaned mechanically. Thus, its application is limited to clean services on the shell side. However, if a satisfactory chemical cleaning program can be employed, fixed-tube sheet construction may be selected for fouling services on the shell side.

In the event of a large differential temperature between the tubes and the shell, the tube sheets will be unable to absorb the differential stress, thereby making it necessary to incorporate an expansion joint. This takes away the advantage of low cost to a significant extent.

U-tube: As the name implies, the tubes of a U-tube heat exchanger are bent in the shape of a U. There is only one tube sheet in a U-tube heat exchanger. However, the lower cost for the single tube sheet is offset by the additional costs incurred for the bending of the tubes and the somewhat larger shell diameter (due to the minimum U-bend radius), making the cost of a U-tube heat exchanger comparable to that of a fixed-tube sheet exchanger.

The advantage of a U-tube heat exchanger is that because one end is free, the bundle can expand or contract in response to stress differentials. In addition, the outsides of the tubes can be cleaned, as the tube bundle can be removed. The disadvantage of the U-tube construction is that the insides of the tubes cannot be cleaned effectively, since the U-bends would require flexible-end drill shafts for cleaning. Thus, U-tube heat exchangers should not be used for services with a dirty fluid inside tubes.

Floating head: The floating-head heat exchanger is the most versatile type of shell and tube heat exchanger (STHE) and also the costliest. In this design, one tube sheet is fixed relative to the shell, and the other is free to "float" within the shell. This permits free expansion of the tube bundle, as well as cleaning of both the insides and outsides of the tubes. Thus, floating-head SHTEs can be used for services where both the shell-side and the tube-side fluids are dirty—making this the standard construction type used in dirty services, such as in petroleum refineries.

There are various types of floating-head construction. The two most common are the pull-through with backing device (TEMA S) and pull-through (TEMA T) designs The TEMA S design (Figure E.5) is the most common configuration in the chemical process industries (CPI). The floating-head cover is secured against the floating tube sheet by bolting it to an ingenious split backing ring. This floating-head closure is located beyond the end of the shell and contained by a shell cover of a larger diameter. To dismantle the heat exchanger, the shell cover is removed first, then the split backing ring, and then the floating-head cover, after which the tube bundle can be removed from the stationary end.

In the TEMA T construction (Figure E.6), the entire tube bundle, including the floating-head assembly, can be removed from the stationary end, since the shell diameter is larger than the floating-head flange. The floating-head cover is bolted directly to the floating tube sheet so that a split backing ring is not required.

The advantage of this construction is that the tube bundle may be removed from the shell without removing either the shell or the floating-head cover, thus reducing maintenance time. This design is particularly suited to kettle reboilers having a dirty heating medium where U-tubes cannot be employed. Due to the enlarged shell, this construction has the highest cost of all exchanger types.

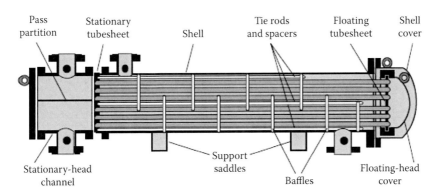

FIGURE E.5
Pull-through floating-head exchanger with backing device (TEMA S).

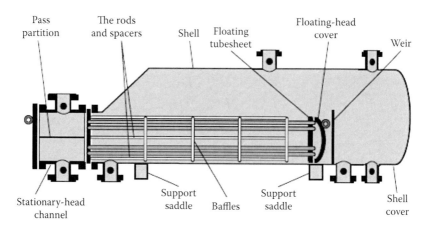

FIGURE E.6
Pull-through floating-head exchanger (TEMA T).

Mechanical Stresses

1. *Sources of mechanical stresses.* Every exchanger is subject to mechanical stresses from a variety of sources in addition to temperature gradients. There are mechanical stresses that result from the construction techniques used on the exchanger, for example, tube and tube sheet stresses resulting from rolling in the tubes. During the manufacture, shipping, and installation of the exchanger, there are many, frequently unforeseen, stresses imposed. There are stresses caused by the support structure reacting to the weight of the exchanger and stresses from the connecting piping; these stresses are generally very different during normal plant operation than during construction or shutdown.

   Finally, these are the stresses arising within the exchanger as a result of the process stream conditions–especially pressure–during operation.

2. *Provision for mechanical stress.* To protect the exchanger from permanent deformation or weakening from these mechanical stresses, it is necessary to design the exchanger so that any stress that can be reasonably expected to occur will not strain or deform the metal beyond the point where it will spontaneously return to its original condition. And it is necessary to ensure that stresses greater than the design values do not occur.

The analysis of stresses and strains in a heat exchanger is an extremely broad and complicated subject and will not be developed in any detail here.

The more obvious problems can be at least anticipated in a qualitative way by the thermal designer who can then seek the advice of a specialist in the subject.

Vibration Problem

A very serious problem in the mechanical design of heat exchangers is flow-induced vibration of the tubes. There are several possible consequences of tube vibration, all of them bad. The tubes may vibrate against the baffles, which can eventually cut holes in the tubes. In extreme cases, the tubes can strike adjacent tubes, literally knocking holes in each other. Or the repeated stressing of the tube near a rigid support such as a tube sheet can result in fatigue cracking of a tube, loosening of the tube joint, and accelerated corrosion.

Vibration is caused by repeated unbalanced forces being applied to the tube. There are a number of such forces, but the most common one in heat exchangers is the eddying motion of the fluid in the wake of a tube as the fluid flows across the tube. The unbalanced forces are relatively small, but they occur tens, hundreds, or thousands of times a second, and their magnitudes increase rapidly with increased fluid velocity. Even so, these forces are ordinarily damped out with no damage to the tube. However, anybody can vibrate much more easily at certain frequencies (called "natural frequencies") than at others. If the unbalanced forces are applied at "driving frequencies" that are at or near these natural frequencies, resonance occurs, and even small forces can result in very strong vibration of the tube.

Erosion

Another essentially mechanical problem in heat exchanger design is that of erosion: the rapid removal of metal due to the friction of the fluid flowing in or across the tube. Erosion often occurs with and accelerates the effect of corrosion by stripping off the protective film formed on certain metals.

The erosion rate depends upon the metal (the harder the metal, the less the erosion if other factors are equal), the velocity and density of the fluid, and the geometry of the system. Thus, erosion is usually more severe at the entrance of a tube or in the bend of a U-tube, due to the additional shear stress associated with developing the boundary layer or turning the fluid. Other more elusive effects are associated with the chemistry of the fluid and the tube metal, especially where corrosion is involved.

There are some commonly used upper velocity limits for flow inside tubes of a given metal. These limits are shown in Table E.1.

Table E.1

Start-Up and Shutdown Procedure of Heat Exchanger

Heat Exchanger Type of Construction	Fluid Location and Relative Temp.				Start-Up Procedure	Shutdown Procedure
	Shell Side		Tube Side			
	Type of Fluid	Rel. Temp.	Type of Fluid	Rel. Temp.		
Fixed-tube sheet (nonremovable bundle)	Liquid	Hot	Liquid	Cold	Start both fluids gradually at the same time.	Shut down both fluids gradually at the same time.
	Condensing gas (e.g., steam)	Hot	Liquid or gas	Cold	Start hot fluid first, and then slowly start cold fluid. Avoid temperature shock (1).	Shut down cold fluid first, then hot fluid.
	Gas	Hot	Liquid	Cold	Start cold fluid first, then hot fluid.	Shut down cold fluid gradually, then hot fluid.
	Liquid	Cold	Liquid	Hot	Start both flows gradually at the same time.	Shut down both fluids gradually at the same time.
	Liquid	Cold	Gas	Hot	Start cold fluid first, then hot fluid.	Shut down hot fluid first, then cold fluid.

	Shell-side fluid		Tube-side fluid		Start-up	Shutdown
U-tube Packed floating head Packed floating tube sheet Internal floating head (All these types have removable bundles)	Liquid	Hot	Liquid	Cold	Start cold fluid first, and then start hot fluid gradually.	Shut down hot fluid first, then cold fluid.
	Condensing gas (e.g., steam)	Hot	Liquid or gas	Cold	Start cold fluid first, and then start hot fluid gradually.	Shut down cold fluid first, and then shut down hot fluid gradually.
	Gas	Hot	Liquid	Cold	Start cold fluid first, and then start hot fluid gradually.	Shut down hot fluid first, then cold fluid.
	Liquid	Cold	Liquid	Hot	Start cold fluid first, and then start hot fluid gradually	Shut down hot fluid first, then cold fluid.
	Liquid	Cold	Gas	Hot	Start cold fluid first, and then start hot fluid gradually.	Shut down hot fluid first, then cold fluid.

General comments

1. In all start-up and shutdown operations, fluid flows should be regulated so as to avoid thermal shocking the unit regardless of whether the unit is of either a removable- or nonremovable-type construction.

2. For fixed-tube sheet (nonremovable bundle)-type units, the tube-side fluid cannot be shut down; it is recommended that (a) a bypass arrangement be incorporated in the system and (b) the tube-side fluid be bypassed before the shell-side fluid is shut down.

3. Extreme caution should be taken on insulated units where fluid flows are terminated and then restarted. Since the metal parts could remain at high temperatures for an extended period, severe thermal shock could occur.

Allocation of Streams in a Shell and Tube Exchanger

In principle, either stream entering a shell and tube exchanger may be put on either side—tube side or shell side—of the surface. However, there are four considerations, which exert a strong influence upon which choice will result in the most economical exchanger:

1. High pressure: If one of the streams is at a high pressure, it is desirable to put that stream inside the tubes. In this case, only the tubes and the tube-side fittings need to be designed to withstand the high pressure, whereas the shell may be made of lighter-weight metal. Obviously, if both streams are at high pressure, a heavy shell will be required and other considerations will dictate which fluid goes in the tube. In any case, high shell-side pressure puts a premium on the design of long, small-diameter exchangers.

2. Corrosion: Corrosion generally dictates the choice of material of construction, rather than exchanger design.

   However, since most corrosion-resistant alloys are more expensive than the ordinary materials of construction, the corrosive fluid will ordinarily be placed in the tubes so that at least the shell need not be made of corrosion-resistant material. If the corrosion cannot be effectively prevented but only slowed by choice of material, a design must be chosen in which corrodible components can be easily replaced (unless it is more economical to scrap the whole unit and start over.)

3. Fouling: Fouling enters into the design of almost every process exchanger to a measurable extent, but certain streams foul so badly that the entire design is dominated by features that seek (a) to minimize fouling (e.g., high velocity, avoidance of dead or eddy flow regions), (b) to facilitate cleaning (fouling fluid on tube-side, wide pitch, and rotated square layout if shell-side fluid is fouling), or (c) to extend operational life by multiple units.

4. Low heat transfer coefficient: If one stream has an inherently low heat transfer coefficient (such as low-pressure gases or viscous liquids), this stream is preferentially put on the shell side so that extended surface may be used to reduce the total cost of the heat exchanger.

Storage

If they cannot be installed and put into operation immediately upon receipt at the jobsite, certain precautions are necessary to prevent deterioration during storage.

Good storage practices are important, considering the high costs of repair or replacement and the possible delays for items, which require long lead times for manufacture. The following suggested practices are provided solely as a convenience to the user, who shall make his or her own decision on whether to use all or any of them:

1. On receipt of the heat exchanger, inspect for shipping damage to all protective covers. If damage is evident, inspect for possible contamination and replace protective covers as required. If damage is extensive, notify the carrier immediately.

2. If the heat exchanger is not to be placed in immediate service, take precautions to prevent rusting or contamination.

3. Heat exchangers for oil service, made of ferrous materials, may be pressure tested with oil at the factory. However, the residual oil coating on the inside surfaces of the exchanger does not preclude the possibility of rust formation. Upon receipt, fill these exchangers with appropriate oil or coat them with a corrosion prevention compound for storage. These heat exchangers have a large warning decal, indicating that they should be protected with oil.

4. The choice of preservation of interior surfaces during storage for other service applications depends upon your system requirements and economics. Only when included in the original purchase order specifications will specific preservation be incorporated prior to shipment from the factory.

5. Remove any accumulations of dirt, water, ice, or snow and wipe dry before moving exchangers into indoor storage. If unit was not filled with oil or other preservatives, open drain plugs to remove any accumulated moisture and then reseal. Accumulation of moisture usually indicates rusting has already started and remedial action should be taken.

6. Store under cover in a heated area, if possible. The ideal storage environment for heat exchangers and accessories is indoors, above grade, and in a dry, low-humidity atmosphere, which is sealed to prevent entry of blowing dust, rain, or snow. Maintain temperatures between 70°F and 105°F (wide temperature swings may cause condensation and "sweating" of steel parts). Cover windows to prevent temperature variations caused by sunlight. Provide thermometers and humidity indicators at several points, and maintain atmosphere at 40% relative humidity or lower.

7. In tropical climates, it may be necessary to use trays of renewable desiccants (such as silica gel), or portable dehumidifiers, to remove moisture from the air in the storage enclosure. Thermostatically controlled portable heaters (vented to outdoors) may be required to maintain even air temperatures inside the enclosure.

8. Inspect heat exchangers and accessories frequently while they are in storage. Start a log to record results of inspections and maintenance performed while units are in storage. A typical log entry should include, for each component, at least the following:
   a. Date
   b. Inspector's name
   c. Identification of unit or item
   d. Location
   e. Condition of paint or coating
   f. Condition of interior
   g. Is free moisture present?
   h. Has dirt accumulated?
   i. Corrective steps taken

9. When paint deterioration begins, as evidenced by discoloration or light rusting, consider touch-up or repainting. If the unit is painted with our standard shop enamel, areas of light rust may be wire brushed and touched-up with any good-quality air-drying synthetic enamel.

   Units painted with special paints (when specified on customers' orders) may require special techniques for touch-up or repair. Obtain specific information from the paint manufacturer. Painted steel units should never be permitted to rust or deteriorate to a point where their strength will be impaired. But a light surface rusting, on steel units, which will be repainted after installation, will not generally cause any harm.

Interiors coated with rust preventive should be restored to good condition and recoated promptly if signs of rust occur.

Installation Planning

1. On removable bundle heat exchangers, provide sufficient clearance at the stationary end to permit the removal of the tube bundle from the shell. On the floating-head end, provide space to permit removal of the shell cover and floating-head cover.

2. On fixed bundle heat exchangers, provide sufficient clearance at one end to permit removal and replacement of tubes and at the other end provide sufficient clearance to permit tube rolling.

3. Provide valves and bypasses in the piping system so that both the shell side and tube side may be bypassed to permit isolation of the heat exchanger for inspection, cleaning, and repairs.

4. Provide convenient means for frequent cleaning as suggested under maintenance.

5. Provide thermometer wells and pressure gauge pipe taps in all piping to and from the heat exchanger, located as close to the heat exchanger as possible.

6. Provide necessary air vent valves for the heat exchanger so that it can be purged to prevent or relieve vapor or gas binding on both the tube side and shell side.

7. Provide adequate supports for mounting the heat exchanger so that it will not settle and cause piping strains. Foundation bolts should be set accurately. In concrete footings, pipe sleeves at least one pipe size larger than the bolt diameter slipped over the bolt and cast in place are best for this purpose as they allow the bolt centers to be adjusted after the foundation has set.

8. Install proper liquid level controls and relief valves and liquid level and temperature alarms.

9. Install gauge glasses or liquid level alarms in all vapor or gas spaces to indicate any failure occurring in the condensate drain system and to prevent flooding of the heat exchanger.

10. Install a surge drum upstream from the heat exchanger to guard against pulsation of fluids caused by pumps, compressors, or other equipment.

11. Do not pipe drain connections to common closed manifold; it makes it more difficult to determine that the exchanger has been thoroughly drained.

Installation at Jobsite

1. If you have maintained the heat exchanger in storage, thoroughly inspect it prior to installation. Make sure it is thoroughly cleaned to remove all preservation materials, unless stored full of the same oil being used in the system, or the coating is soluble in the lubricating system oil.

2. If the heat exchanger isn't being stored, inspect for shipping damage to all protective covers upon receipt at the jobsite. If damage is evident, inspect for possible contamination and replace protective covers as required. If damage is extensive, notify the carrier immediately.

3. When installing, set heat exchanger level and square so that pipe connections can be made without forcing.

4. Before piping up, inspect all openings in the heat exchanger for foreign material. Remove all wooden plugs, bags of desiccants, and

shipping covers immediately prior to installing. Do not expose internal passages of the heat exchanger to the atmosphere since moisture or harmful contaminants may enter the unit and cause severe damage to the system due to freezing and/or corrosion.

5. After piping is complete, if support cradles or feet are fixed to the heat exchanger, loosen foundation bolts at one end of the exchanger to allow free movement. Oversized holes in support cradles or feet are provided for this purpose.

6. If heat exchanger shell is equipped with a bellow-type expansion joint, remove shipping supports per instructions.

Operation

Be sure entire system is clean before starting operation to prevent plugging of tubes or shell-side passages with refuse.

The use of strainers or settling tanks in pipelines leading to the heat exchanger is recommended.

Open vent connections before starting up.

Start operating gradually. See Table E.1 for suggested start-up and shut-down procedures for most applications.

After the system is completely filled with the operating fluids and all air has been vented, close all manual vent connections.

Retighten bolting on all gasketed or packed joints after the heat exchanger has reached operating temperatures to prevent leaks and gasket failures. Standard published torque values do not apply to packed end joints.

Do not operate the heat exchanger under pressure and temperature conditions in excess of those specified on the nameplate.

To guard against water hammer, drain condensate from steam heat exchangers and similar apparatus when both starting up and shutting down.

Drain all fluids when shutting down to eliminate possible freezing and corroding.

In all installations, there should be no pulsation of fluids, since this causes vibration and will result in reduced operating life.

Under no circumstances is the heat exchanger to be operated at a flow rate greater than that shown on the design specifications. Excessive flows can cause vibration and severely damage the heat exchanger tube bundle.

Heat exchangers that are out of service for extended periods of time should be protected against corrosion as described in the storage requirements for new heat exchangers.

Heat exchangers that are out of service for short periods and use water as the flowing medium should be thoroughly drained and blown dry with warm air, if possible. If this is not practical, the water should be circulated through the heat exchanger on a daily basis to prevent stagnant water conditions that can ultimately precipitate corrosion.

Maintenance

1. Clean exchangers subject to fouling (scale, sludge deposits, etc.) periodically, depending on specific conditions. A light sludge or scale coating on either side of the tube greatly reduces its effectiveness. A marked increase in pressure drop and/or reduction in performance usually indicates cleaning is necessary. Since the difficulty of cleaning increases rapidly as the scale thickens or deposits increase, the intervals between cleanings should not be excessive.

2. Neglecting to keep tubes clean may result in random tube plugging. Consequent overheating or cooling of the plugged tubes, as compared to surrounding tubes, will cause physical damage and leaking tubes due to differential thermal expansion of the metals.

3. To clean or inspect the inside of the tubes, remove only the necessary tube-side channel covers or bonnets, depending on type of exchanger construction.

4. If the heat exchanger is equipped with sacrificial anodes or plates, replace these as required.

5. To clean or inspect the outside of the tubes, it may be necessary to remove the tube bundle. (Fixed-tube sheet exchanger bundles are nonremovable.)

6. When removing tube bundles from heat exchangers for inspection or cleaning, exercise care to see that they are not damaged by improper handling:
    - The weight of the tube bundle should not be supported on individual tubes but should be carried by the tube sheets, support or baffle plates, or on blocks contoured to the periphery of the tube bundles.
    - Do not handle tube bundles with hooks or other tools, which might damage tubes. Move tube bundles on cradles or skids.
    - To withdraw tube bundles, pass rods through two or more of the tubes and take the load on the floating tube sheet.
    - Rods should be threaded at both ends, provided with nuts, and should pass through a steel bearing plate at each end of the bundle.
    - Insert a soft wood filler board between the bearing plate and tube sheet face to prevent damage to the tube ends.

- Screw forged steel eyebolts into both bearing plates for pulling and lifting.
- As an alternate to the rods, thread a steel cable through one tube and return through another tube.
- A hardwood spreader block must be inserted between the cable and each tube sheet to prevent damage to the tube ends.

7. If the heat exchanger has been in service for a considerable length of time without being removed, it may be necessary to use a jack on the floating tube sheet to break the bundle free:

   - Use good-sized steel bearing plate with a filler board between the tube sheet face and bearing plate to protect the tube ends.

8. Lift tube bundles horizontally by means of a cradle formed by bending a light gauge plate or plates into a U shape. Make attachments in the legs of the U for lifting.

9. Do not drag bundles, since baffles or support plates may become easily bent. Avoid any damage to baffles so that the heat exchanger will function properly.

10. Some suggested methods of cleaning either the shell side or tube side are listed in the following:

    - Circulating hot wash oil or light distillate through tube side or shell side will usually effectively remove sludge or similar soft deposits.
    - Soft salt deposits may be washed out by circulating hot fresh water.
    - Some commercial cleaning compounds such as "Oakite" or "Dowell" may be effective in removing more stubborn deposits. Use in accordance with the manufacturer's instructions.

11. Table E.2 shows safe loads for steel rods and eyebolts.

12. To locate ruptured or corroded tubes or leaking joints between tubes and tube sheets, the following procedure is recommended.

Table E.2

Safe Loads for Steel Rods and Eyebolts

Steel Rods			Steel Eyebolts	
Tube Size (in.)	Rod Size (in.)	Safe Load per Rod (lbs)	Size (in.)	Safe Load (lbs)
5/8	3/8	1000	3/4	4,000
3/4	1/2	2000	1	6,000
1 or	—	—	1–1/4	10,000
Larger	5/8	3000	1–1/2	15,000

13. To locate ruptured or corroded tubes or leaking joints between tubes and tube sheets, the following procedure is recommended:
    - Remove tube-side channel covers or bonnets.
    - Pressurize the shell side of the exchanger with a cold fluid, preferably water.
    - Observe tube joints and tube ends for indication of test fluid leakage.
14. With certain styles of exchangers, it will be necessary to buy or make a test ring to seal off the space between the floating tube sheet and inside shell diameter to apply the test.
15. To tighten a leaking tube joint, use a suitable parallel roller tube expander:
    - Do not roll tubes beyond the back face of the tube sheet. Maximum rolling depth should be tube sheet thickness −1/8 in.
    - Do not reroll tubes that are not leaking since this needlessly thins the tube wall.
16. It is recommended that when a heat exchanger is dismantled, new gaskets should be used in reassembly:
    - Composition gaskets become brittle and dried out in service and do not provide an effective seal when reused.
    - Metal or metal-jacketed gaskets in initial compression match the contact surfaces and tend to work-harden and cannot be recompressed on reuse.
17. Use of new bolting in conformance with dimension and ASTM specifications of the original design is recommended where frequent dismantling is encountered.

Fouling in Heat Exchangers

"Fouling" is a general term that includes any kind of deposit of extraneous material that appears upon the heat transfer surface during the lifetime of the heat exchanger. Whatever the cause or exact nature of the deposit, additional resistances to heat transfer are introduced and the operational capability of the heat exchanger is correspondingly reduced. In many cases, the deposit is heavy enough to significantly interfere with fluid flow and increase the pressure drop required to maintain the flow rate through the exchanger.

The designer must consider the effect of fouling upon heat exchanger performance during the desired operational lifetime and make provisions in his or her design for sufficient extra capacity to ensure that the exchanger will meet process specifications up to shut down for cleaning. The designer must also consider what mechanical arrangements are necessary to permit easy cleaning.

Types of Fouling

There are several different basic mechanisms by which fouling deposits may be created and each of them in general depends upon several variables. In addition, two or more fouling mechanisms can occur in conjunction in a given service. In this section, we will identify the major mechanisms of fouling and the more important variables upon which they depend:

1. *Sedimentation fouling.* Many streams and particularly cooling water contain suspended solids, which can settle out upon the heat transfer surface. Usually the deposits thus formed do not adhere strongly to the surface and are self-limiting; that is, the thicker a deposit becomes, the more likely it is to wash off (in patches) and thus attain some asymptotic average value over a period of time. Sedimentation fouling is strongly affected by velocity and less so by wall temperature. However, a deposit can "bake on" to a hot wall and become very difficult to remove.

2. *Inverse solubility fouling.* Certain salts commonly found in natural waters—notably calcium sulfate—are less soluble in warm water than in cold. If such a stream encounters a wall at a temperature above that corresponding to saturation for the dissolved salt, the salt will crystallize on the surface. Crystallization will begin at special active points—nucleation sites—such as scratches and pits, often after a considerable induction period, and then spread to cover the entire surface. The buildup will continue as long as the surface in contact with the fluid has a temperature above saturation. The scale is strong and adherent and usually requires vigorous mechanical or chemical treatment to remove it.

3. *Chemical reaction fouling.* The previously mentioned fouling mechanisms involve primarily physical changes. Common sources of fouling on the process stream side are chemical reactions that result in producing a solid phase at or near the surface. For example, a hot heat transfer surface may cause thermal degradation of one of the components of a process stream, resulting in carbonaceous deposits (commonly called "coke") on the surface. Or a surface may cause polymerization to occur, resulting in a tough layer of low-grade plastic or synthetic rubber. These deposits are often extremely tenacious and may require such extreme measures as burning off the deposit in order to return the exchanger to satisfactory operation.

4. *Corrosion product fouling.* If a stream corrodes the metal of the heat transfer surface, the corrosion products may be essential to protect the remaining metal against further corrosion, in which case any attempt to clean the surface may only result in accelerated corrosion and failure of the exchanger.

5. *Biological fouling.* Many cooling water sources and a few process streams contain organisms that will attach to solid surfaces and grow. These organisms range from algae and microbial slimes to barnacles and mussels. Even when only a very thin film is present, the heat transfer resistance can be very great. Where macroscopic forms like mussels are present, the problem is no longer one of heat transfer—there won't be any through the animal—but rather of plugging up the flow channels. If biological fouling is thought to be a problem, the usual solution is to kill the life forms by chlorination or to discourage their settling on the heat transfer surface by using 90–10 copper–nickel (alloy C70600) or other high copper alloy tubes. As an alternative to continuous chlorination, intermittent "shock" chlorination may be successful.

6. Combined Mechanisms. Most of the earlier fouling processes can occur in combination. A common example is the combination of (1) and (2) in cooling tower water. Most surface waters contain both sediment and calcium carbonates, and the concentrations of these components rise as the water is recirculated through the cooling system.

It is therefore common to find deposits composed of crystals of inverse solubility salts together with finely divided sediments. The behavior of these deposits is intermediate between the two limiting cases: The crystals tend to hold the sediment in place, but there are planes of weakness in the structure that fail from time to time and cause the deposit to break off in patches.

Effect of Fouling on Heat Transfer

As noted earlier, the effect of fouling is to form an essentially solid deposit upon the surface, through which heat must be transferred by conduction. If we knew both the thickness and the thermal conductivity of the fouling, we could treat the heat transfer problem simply as another conduction resistance in series with the wall. In general, we know neither of these quantities, and the only possible technique is to introduce the additional resistance as fouling factors in computing the overall heat transfer coefficient as previously discussed.

Fouling effects inside the tube usually cause no particular problems if allowance has been made for the reduction in heat transfer and the small increase in flow resistance. However, fouling on the outside of finned tubes can be a more complicated matter, because in extreme situations, there is a possibility that the finite thickness of the fouling layer can effectively close off the flow through the fins. On the other hand, finned surfaces are sometimes found to be more resistant to fouling than plain surfaces. The reasons for this are not well established, though it may be that the expansion and contraction of the surface during normal operational cycles tend to break off brittle fouling films. Caution is advised, however, in applying finned tubes to services known to be significantly fouling.

High-finned tubes are commonly used only with air and other low-pressure and relatively clean gases. Such fouling as does occur is mostly dust deposition, which can easily be removed by blowing.

Material Selection for Fouling Services

Potential fouling problems may influence material selection in one of three ways. Most obvious perhaps is the minimization of corrosion-type fouling by choosing a material of construction, which does not readily corrode or produce voluminous deposits of corrosion products. If chemical removal of the fouling deposit is planned, the material selected must also be resistant to attack by the cleaning solutions.

Secondly, biological fouling can be largely eliminated by the selection of copper-bearing alloys, such as 90–10 copper–nickel (UNS 70600) or 70–30 copper–nickel (UNS 71500). Generally, alloys containing copper in quantities greater than 70% are effective in preventing or minimizing biological fouling.

Thirdly, some types of fouling can be controlled or minimized by using high flow velocities. If this technique is to be employed, the possibility of metal erosion should be considered as it is important to restrict the velocity and/or its duration to values consistent with satisfactory tube life. Some metals, such as titanium or stainless steels, can be quite resistant to erosion by the high-velocity effluent. Recently, a new copper–nickel alloy containing nominally 83% copper, 17% nickel, and 0.4% chromium has been developed by the International Nickel Company (alloy C72200). This alloy is similar to the other copper–nickels relative to its corrosion resistance but, however, has a much greater resistance to velocity attack, being capable of operating in seawater at velocities approaching 25 ft/s.

Removal of Fouling

If fouling cannot be prevented from forming, it is necessary to make some provision for its periodic removal. Some deposits can be removed by purely chemical means; for example, removal of carbonate deposits by chlorination. The application of chemical cleaning techniques is a specialized art and should be undertaken only under the guidance of a specialist. However, since chemical cleaning ordinarily does not require removal of equipment or disassembly of the piping (if properly designed), it is the most convenient of the cleaning techniques in those cases where it can be used.

There are a number of techniques for mechanical removal of fouling. Scraping or rotary brushing is limited to those surfaces that can be reached by the scraping tool—a problem that is eased on the shell side by the use of large clearances between tubes and/or the use of rotated square tube layout. (It should be noted that scraping should not be used on finned tubes.) Use of very high-velocity water jets is very common both inside and outside the tubes, though for the shell side the jets will not be very effective deep inside a large tube bank.

For situations where there is a high premium for maintaining a high degree of cleanliness, for example, large power plant condensers, it may be possible to install a system for the continuous onstream cleaning of the interior surfaces of the tube. The Amertap® system utilizes slightly oversized, sponge rubber balls, which are continuously recirculated through the tubes on a random basis. As the balls pass through the tubes, they remove the accumulation of scale or corrosion products. A mesh basket in the outlet piping collects the balls and a ball pump reinjects them into the entering water flow. The type of balls having an abrasive secured to the outer surface should be used only with great caution as continuous abrasive action may shorten tube life due to the removal of the protective corrosion film formed in copper and copper-base alloys.

Appendix F: Dimension of Flanges and Fittings

Properties of Steel Tubing

OD of Tubing (in.)	Wall Thickness (in.)	Internal Area (sq. in.)	External Surface per Foot Length (sq. ft.)	Internal Surface per Foot Length (sq. ft.)	Theoretical Weight per Foot Length	ID Tubing (in.)	Constants (C*)	OD/ID	Metal Area (Transverse Metal Area) (sq. in.)
5/8	0.125	0.1104	0.1636	0.0982	0.668	0.375	172	1.667	0.1964
5/8	0.110	0.1288	0.1636	0.1060	0.605	0.405	201	1.543	0.1780
5/8	0.105	0.1353	0.1636	0.1086	0.583	0.415	211	1.506	0.1715
5/8	0.095	0.1486	0.1636	0.1139	0.538	0.435	232	1.437	0.1582
5/8	0.085	0.1626	0.1636	0.1191	0.490	0.455	254	1.374	0.1442
5/8	0.075	0.1772	0.1636	0.1244	0.441	0.475	276	1.316	0.1296
5/8	0.065	0.1924	0.1636	0.1296	0.389	0.495	300	1.263	0.1144
5/8	0.060	0.2003	0.1636	0.1322	0.362	0.505	312	1.238	0.1065
5/8	0.055	0.2083	0.1636	0.1348	0.335	0.515	325	1.214	0.0985
5/8	0.050	0.2165	0.1636	0.1374	0.307	0.525	338	1.190	0.0903
3/4	0.150	0.1590	0.1963	0.1178	0.961	0.450	248	1.667	0.2827
3/4	0.135	0.1810	0.1963	0.1257	0.887	0.480	282	1.563	0.2608
3/4	0.125	0.1964	0.1963	0.1309	0.834	0.500	306	1.500	0.2454
3/4	0.110	0.2206	0.1963	0.1388	0.752	0.530	344	1.415	0.2212
3/4	0.105	0.2290	0.1963	0.1414	0.723	0.540	357	1.389	0.2128
3/4	0.095	0.2463	0.1963	0.1466	0.665	0.560	384	1.339	0.1955
3/4	0.085	0.2642	0.1963	0.1518	0.604	0.580	412	1.293	0.1776
3/4	0.075	0.2827	0.1963	0.1571	0.541	0.600	441	1.250	0.1590
3/4	0.065	0.3019	0.1963	0.1623	0.476	0.620	471	1.210	0.1399
3/4	0.060	0.3117	0.1963	0.1649	0.442	0.630	486	1.190	0.1301
3/4	0.055	0.3217	0.1963	0.1676	0.408	0.640	502	1.172	0.1201
3/4	0.050	0.3318	0.1963	0.1702	0.374	0.650	518	1.154	0.1100

7/8	0.150	0.2597	0.2291	0.1505	1.161	0.575	405	1.522	0.3416
7/8	0.135	0.2875	0.2291	0.1584	1.067	0.605	448	1.446	0.3138
7/8	0.125	0.3068	0.2291	0.1636	1.001	0.625	478	1.400	0.2945
7/8	0.110	0.3370	0.2291	0.1715	0.899	0.655	526	1.336	0.2644
7/8	0.105	0.3473	0.2291	0.1741	0.863	0.665	542	1.316	0.2540
7/8	0.095	0.3685	0.2291	0.1793	0.791	0.685	575	1.277	0.2328
7/8	0.085	0.3904	0.2291	0.1846	0.717	0.705	609	1.241	0.2110
7/8	0.075	0.4128	0.2291	0.1898	0.641	0.725	644	1.207	0.1885
7/8	0.065	0.4359	0.2291	0.1950	0.562	0.745	680	1.174	0.1654
7/8	0.060	0.4477	0.2291	0.1977	0.522	0.755	698	1.159	0.1536
7/8	0.055	0.4596	0.2291	0.2003	0.482	0.765	717	1.144	0.1417
7/8	0.050	0.4717	0.2291	0.2029	0.441	0.775	736	1.129	0.1296
1	0.150	0.3848	0.2618	0.1833	1.362	0.700	600	1.429	0.4006
1	0.135	0.4185	0.2618	0.1911	1.247	0.730	653	1.370	0.3669
1	0.125	0.4418	0.2618	0.1964	1.168	0.750	689	1.333	0.3436
1	0.110	0.4778	0.2618	0.2042	1.046	0.790	745	1.282	0.3076
1	0.105	0.4902	0.2618	0.2068	1.004	0.790	764	1.266	0.2952
1	0.095	0.5153	0.2618	0.2121	0.918	0.810	804	1.235	0.2701
1	0.085	0.5411	0.2618	0.2173	0.831	0.830	844	1.205	0.2443
1	0.075	0.5675	0.2618	0.2225	0.741	0.850	885	1.176	0.2179
1	0.065	0.5945	0.2618	0.2278	0.649	0.870	927	1.149	0.1909
1	0.060	0.6082	0.2618	0.2304	0.602	0.880	949	1.136	0.1772
1	0.055	0.6221	0.2618	0.2330	0.555	0.890	970	1.124	0.1633
1	0.050	0.6362	0.2618	0.2356	0.507	0.900	992	1.111	0.1492

150 lb Flanges
Standard ANSI B16.5

1. All dimensions are in inches.
2. Material most commonly used, forged steel SA 105. Available also in stainless steel, alloy steel, and nonferrous metal.
3. The 1/16 in. raised face is included in dimensions C, D, and J.
4. The lengths of stud bolts do not include the height of crown.
5. Bolt holes are 1/8 in. larger than bolt diameters.
6. Flanges bored to dimensions shown unless otherwise specified.
7. Flanges for pipe sizes 22, 26, 28, and 30 are not covered by ANSI B16.5

See Facing Page for Dimension K and Data on Bolting.

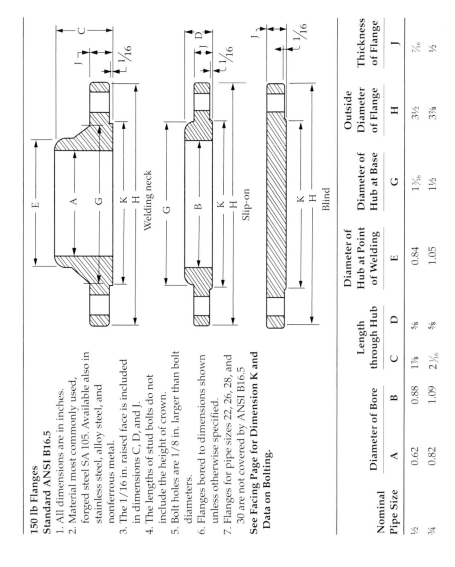

Nominal Pipe Size	Diameter of Bore		Length through Hub		Diameter of Hub at Point of Welding	Diameter of Hub at Base	Outside Diameter of Flange	Thickness of Flange
	A	B	C	D	E	G	H	J
½	0.62	0.88	1⅞	⅝	0.84	1 3⁄16	3½	7⁄16
¾	0.82	1.09	2 1⁄16	⅝	1.05	1½	3⅞	½

1	1.05	1.36	2 3/16	11/16	1.32	1 15/16	4¼	9/16
1¼	1.38	1.70	2¼	13/16	1.66	2 5/16	4⅝	⅝
1½	1.61	1.95	2 7/16	⅞	1.90	2 9/16	5	11/16
2	2.07	2.44	2½	1	2.38	3 3/16	6	¾
2½	2.47	2.94	2¾	1½	2.88	3 9/16	7	⅞
3	3.07	3.57	2¾	1 3/16	3.50	4¼	7½	15/16
3½	3.55	4.07	2 13/16	1¼	4.00	4 13/16	8½	15/16
4	4.03	4.57	3	1 5/16	4.50	5 5/16	9	15/16
5	5.05	5.66	3½	1 7/16	5.56	6 7/16	10	15/16
6	6.07	6.72	3½	1 9/16	6.63	7 9/16	11	1
8	7.98	8.72	4	1¾	8.63	9 11/16	13½	1⅛
10	10.02	10.88	4	1 15/16	10.75	12	16	1 3/16
12	12.00	12.88	4½	2 3/16	12.75	14⅜	19	1¼
14	13.25	14.14	5	2¼	14.00	15¾	21	1⅜
16	15.25	16.16	5	2½	16.00	18	23½	1 7/16
18	17.25	18.18	5½	2 11/16	18.00	19⅞	25	1 9/16
20	19.25	20.20	5 11/16	2⅞	20.00	22	27½	1 11/16
22	21.25	22.22	5⅞	3⅛	22.00	24¼	29½	1 13/16
24	23.25	24.25	6	3¼	24.00	26⅛	32	1⅞
26	To be specified	26.25	5	3⅜	26.00	28½	34¼	2
28		28.25	5 1/16	3 7/16	28.00	30¾	36½	2 1/16
30		30.25	5⅛	3½	30.00	32¾	38¾	2⅛

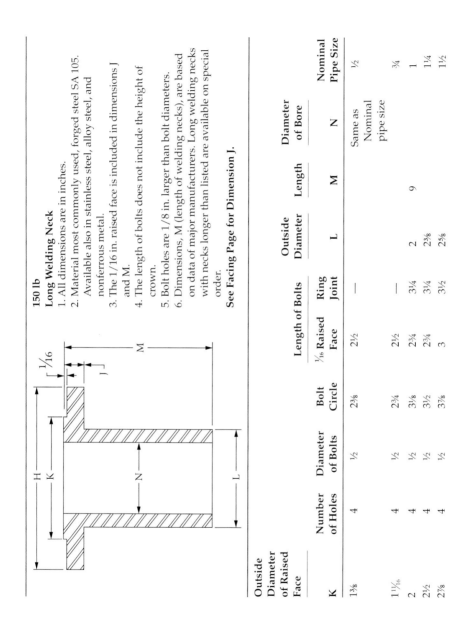

150 lb
Long Welding Neck
1. All dimensions are in inches.
2. Material most commonly used, forged steel SA 105. Available also in stainless steel, alloy steel, and nonferrous metal.
3. The 1/16 in. raised face is included in dimensions J and M.
4. The length of bolts does not include the height of crown.
5. Bolt holes are 1/8 in. larger than bolt diameters.
6. Dimensions, M (length of welding necks), are based on data of major manufacturers. Long welding necks with necks longer than listed are available on special order.

See Facing Page for Dimension J.

Outside Diameter of Raised Face K	Number of Holes	Diameter of Bolts	Bolt Circle	Length of Bolts		Outside Diameter L	Length M	Diameter of Bore N	Nominal Pipe Size
				1/16 Raised Face	Ring Joint				
1 3/8	4	1/2	2 3/8	2 1/2	—			Same as Nominal pipe size	1/2
1 11/16	4	1/2	2 3/4	2 1/2	—	2	9		3/4
2	4	1/2	3 1/8	2 3/4	3 1/4	2 3/8			1
2 1/2	4	1/2	3 1/2	2 3/4	3 1/4	2 5/8			1 1/4
2 7/8	4	1/2	3 7/8	3	3 1/2				1 1/2

35/8	4	5/8	43/4	31/4	33/4	31/4		2
41/8	4	5/8	51/2	31/2	4	33/4		21/2
5	4	5/8	6	33/4	41/4	41/4		3
51/2	8	5/8	7	33/4	41/4	47/8		31/2
6 3/16	8	5/8	71/2	33/4	41/4	51/2	12	4
7 5/16	8	3/4	81/2	4	41/2	61/2		5
81/2	8	3/4	91/2	4	41/2	73/4		6
105/8	8	3/4	113/4	41/4	43/4	93/4		8
123/4	12	7/8	141/4	43/4	51/4	12		10
15	12	7/8	17	43/4	51/4	143/8		12
161/4	12	1	183/4	51/4	53/4	16	10–14	14
181/2	16	1	211/4	51/2	6	18		16
21	16	11/8	223/4	6	61/2	20		18
23	20	11/8	25	61/4	63/4	22		20
	20	11/4	271/4	61/2	7			22
271/4	20	11/4	291/2	7	71/2	261/4		24
291/4	24	11/4	313/4	7	—	281/2		26
311/4	28	11/4	34	7	—	301/2		28
333/4	28	11/4	36	71/4	—	321/2		30

300 lb Flanges
Standard ANSI B16.5

1. All dimensions are in inches.
2. Material most commonly used, forged steel SA 105. Available also in stainless steel, alloy steel, and nonferrous metal.
3. The 1/16 in. raised face is included in dimensions C, D, and J.
4. The lengths of stud bolts do not include the height of crown.
5. Bolt holes are 1/8 in. larger than bolt diameters.
6. Flanges bored to dimensions shown unless otherwise specified.
7. Flanges for pipe sizes 22, 26, 28, and 30 are not covered by ANSI B16.5.

See Facing Page for Dimension K and Data on Bolting.

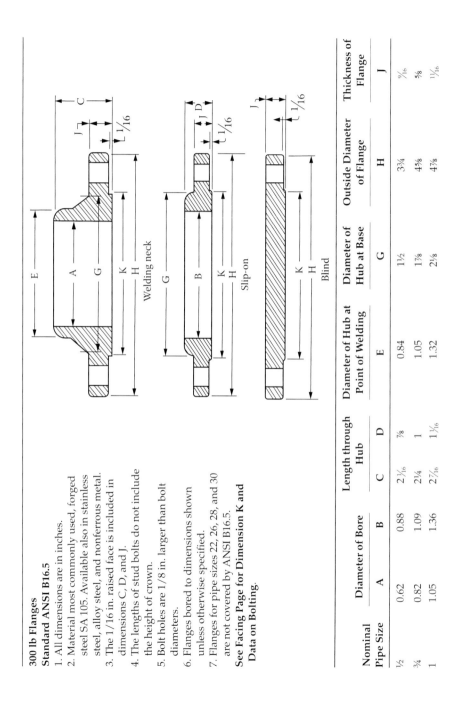

Welding neck

Slip-on

Blind

Nominal Pipe Size	Diameter of Bore		Length through Hub		Diameter of Hub at Point of Welding	Diameter of Hub at Base	Outside Diameter of Flange	Thickness of Flange
	A	B	C	D	E	G	H	J
½	0.62	0.88	2 1/16	7/8	0.84	1½	3¾	9/16
¾	0.82	1.09	2¼	1	1.05	1⅞	4⅝	5/8
1	1.05	1.36	2 7/16	1 1/16	1.32	2⅛	4⅞	11/16

1¼	1.38	1.70	2 9/16	1 1/16	1.66	2½	5¼	¾
1½	1.61	1.95	2 11/16	1 3/16	1.90	2¾	6⅛	13/16
2	2.07	2.44	2¾	1 5/16	2.38	3 5/16	6½	⅞
2½	2.47	2.94	3	1½	2.88	3 15/16	7½	1
3	3.07	3.57	3⅛	1 11/16	3.50	4⅝	8¼	1⅛
3½	3.55	4.07	3 3/16	1¾	4.00	5¼	9	1 3/16
4	4.03	4.57	3⅜	1⅞	4.50	5¾	10	1¼
5	5.05	5.66	3⅞	2	5.56	7	11	1⅜
6	6.07	6.72	3⅞	2 1/16	6.63	8⅛	12½	1 7/16
8	7.98	8.72	4⅜	2 7/16	8.63	10¼	15	1⅝
10	10.02	10.88	4⅝	2⅝	10.75	12⅝	17½	1⅞
12	12.00	12.88	5⅛	2⅞	12.75	14¾	20½	2
14	13.25	14.14	5⅝	3	14.00	16¾	23	2⅛
16	15.25	16.16	5¾	3¼	16.00	19	25½	2¼
18	17.25	18.18	6¼	3½	18.00	21	28	2⅜
20	19.25	20.20	6⅜	3¾	20.00	23⅛	30½	2½
22	21.25	22.22	6½	4	22.00	25¼	33	2⅝
24	23.25	24.25	6⅝	4 3/16	24.00	27⅝	36	2¾
26	To be specified	26.25	7¼	7¼	26¼	28⅜	38¼	3⅛
28		28.25	7¾	7¾	28¼	30½	40¾	3⅜
30		30.25	8¼	8¼	30¼	32 9/16	43	3⅝

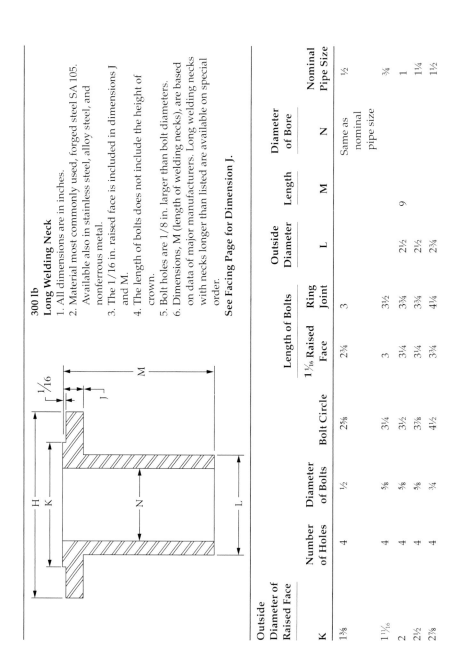

300 lb

Long Welding Neck

1. All dimensions are in inches.
2. Material most commonly used, forged steel SA 105. Available also in stainless steel, alloy steel, and nonferrous metal.
3. The 1/16 in. raised face is included in dimensions J and M.
4. The length of bolts does not include the height of crown.
5. Bolt holes are 1/8 in. larger than bolt diameters.
6. Dimensions, M (length of welding necks), are based on data of major manufacturers. Long welding necks with necks longer than listed are available on special order.

See Facing Page for Dimension J.

| Outside Diameter of Raised Face | Number of Holes | Diameter of Bolts | Bolt Circle | Length of Bolts | | Outside Diameter | | | Diameter of Bore | Nominal Pipe Size |
K				1 7/16 Raised Face	Ring Joint		L	M	N	
1⅜	4	½	2⅝	2¾	3				Same as nominal pipe size	½
1 11/16	4	⅝	3¼	3	3½					¾
2	4	⅝	3½	3¼	3¾		2½	9		1
2½	4	⅝	3⅞	3¼	3¾		2½			1¼
2⅞	4	¾	4½	3¾	4¼		2¾			1½

2			$3\frac{3}{16}$	$4\frac{1}{4}$	$3\frac{1}{2}$	5	$\frac{5}{8}$	8	$3\frac{5}{8}$
$2\frac{1}{2}$			$3\frac{15}{16}$	$4\frac{3}{4}$	4	$5\frac{7}{8}$	$\frac{3}{4}$	8	$4\frac{1}{4}$
3			$4\frac{5}{8}$	5	$4\frac{1}{4}$	$6\frac{5}{8}$	$\frac{3}{4}$	8	5
$3\frac{1}{2}$	12		$5\frac{1}{4}$	$5\frac{1}{4}$	$4\frac{1}{2}$	$7\frac{1}{4}$	$\frac{3}{4}$	8	$5\frac{1}{2}$
4			$5\frac{3}{4}$	$5\frac{1}{4}$	$4\frac{1}{2}$	$7\frac{7}{8}$	$\frac{3}{4}$	8	$6\frac{3}{16}$
5			7	$5\frac{1}{2}$	$4\frac{3}{4}$	$9\frac{1}{4}$	$\frac{3}{4}$	8	$7\frac{5}{16}$
6			$8\frac{1}{8}$	$5\frac{3}{4}$	5	$10\frac{5}{8}$	$\frac{3}{4}$	12	$8\frac{1}{2}$
8			$10\frac{1}{4}$	$6\frac{1}{4}$	$5\frac{1}{2}$	13	$\frac{7}{8}$	12	$10\frac{5}{8}$
10			$12\frac{5}{8}$	7	$6\frac{1}{4}$	$15\frac{1}{4}$	1	16	$12\frac{3}{4}$
12			$14\frac{3}{4}$	$7\frac{1}{2}$	$6\frac{3}{4}$	$17\frac{3}{4}$	$1\frac{1}{8}$	16	15
14		10–14	$16\frac{3}{4}$	$7\frac{3}{4}$	7	$20\frac{1}{4}$	$1\frac{1}{8}$	20	$16\frac{1}{4}$
16			19	$8\frac{1}{4}$	$7\frac{1}{2}$	$22\frac{1}{2}$	$1\frac{1}{4}$	20	$18\frac{1}{2}$
18			21	$8\frac{1}{2}$	$7\frac{3}{4}$	$24\frac{3}{4}$	$1\frac{1}{4}$	24	21
20			$23\frac{1}{8}$	9	$8\frac{1}{4}$	27	$1\frac{1}{4}$	24	23
22				$9\frac{3}{4}$	$8\frac{3}{4}$	$29\frac{1}{4}$	$1\frac{1}{2}$	24	$25\frac{1}{4}$
24			$27\frac{5}{8}$	$10\frac{1}{4}$	$9\frac{3}{4}$	32	$1\frac{1}{2}$	24	$27\frac{1}{4}$
26			$29\frac{1}{2}$	11	10	$34\frac{1}{2}$	$1\frac{5}{8}$	28	$29\frac{1}{2}$
28			$31\frac{1}{2}$	$11\frac{1}{2}$	$10\frac{1}{2}$	37	$1\frac{5}{8}$	28	$31\frac{1}{2}$
30			$33\frac{3}{4}$	$12\frac{1}{4}$	$11\frac{1}{4}$	$39\frac{1}{4}$	$1\frac{3}{4}$	28	$33\frac{3}{4}$

400 lb Flanges
Standard ANSI B16.5

1. All dimensions are in inches.
2. Material most commonly used, forged steel SA 105. Available also in stainless steel, alloy steel, and nonferrous metal.
3. The 1/4 in. raised face is not included in dimensions C, D, and J.
4. The lengths of stud bolts do not include the height of crown.
5. Bolt holes are 1/8 in. larger than bolt diameters.
6. Flanges bored to dimensions shown unless otherwise specified.
7. Flanges for pipe sizes 22, 26, 28, and 30 are not covered by ANSI B16.5.

See Facing Page for Dimension K and Data on Bolting.

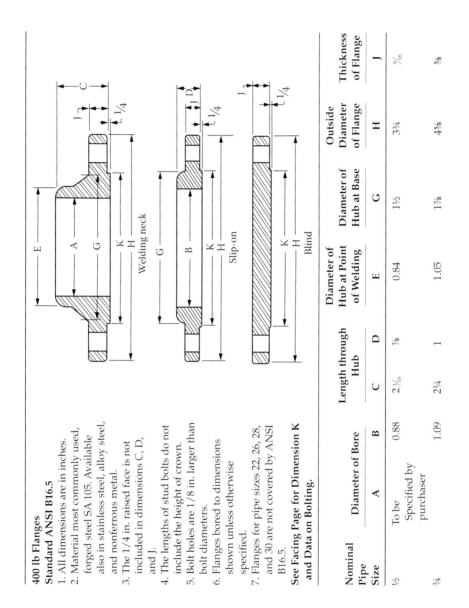

Welding neck

Slip-on

Blind

Nominal Pipe Size	Diameter of Bore		Length through Hub		Diameter of Hub at Point of Welding	Diameter of Hub at Base	Outside Diameter of Flange	Thickness of Flange
	A	B	C	D	E	G	H	J
½	To be Specified by purchaser	0.88	2 1/16	7/8	0.84	1½	3¾	9/16
¾		1.09	2¼	1	1.05	1⅞	4⅝	⅝

1	1.36	2 7/16	1 1/16	1.32	2 1/8	4 7/8	11/16
1 1/4	1.70	2 5/8	1 1/8	1.66	2 1/2	5 1/4	13/16
1 1/2	1.95	2 3/4	1 1/4	1.90	2 3/4	6 1/8	7/8
2	2.44	2 7/8	1 7/16	2.38	3 5/16	6 1/2	1
2 1/2	2.94	3 1/8	1 5/8	2.88	3 15/16	7 1/2	1 1/8
3	3.57	3 1/4	1 13/16	3.50	4 5/8	8 1/4	1 1/4
3 1/2	4.07	3 3/8	1 15/16	4.00	5 1/4	9	1 3/8
4	4.57	3 1/2	2	4.50	5 3/4	10	1 3/8
5	5.66	4	2 1/8	5.56	7	11	1 1/2
6	6.72	4 1/16	2 1/4	6.63	8 1/8	12 1/2	1 5/8
8	8.72	4 5/8	2 11/16	8.63	10 1/4	15	1 7/8
10	10.88	4 7/8	2 7/8	10.75	12 5/8	17 1/2	2 1/8
12	12.88	5 3/8	3 1/8	12.75	14 3/4	20 1/2	2 1/4
14	14.14	5 7/8	3 5/16	14.00	16 3/4	23	2 3/8
16	16.16	6	3 11/16	16.00	19	25 1/2	2 1/2
18	18.18	6 1/2	3 7/8	18.00	21	28	2 5/8
20	20.20	6 5/8	4	20.00	23 1/8	30 1/2	2 3/4
22	22.22	6 3/4	4 1/4	22.00	25 1/4	33	2 7/8
24	24.25	6 7/8	4 1/2	24.00	27 5/8	36	3
26	26.25	7 5/8	7 5/8	26 5/16	28 5/8	38 1/4	3 1/2
28	28.25	8 1/8	8 1/8	28 5/16	30 13/16	40 3/4	3 3/4
30	30.25	8 5/8	8 5/8	30 5/16	32 15/16	43	4

400 lb
Long Welding Neck

1. All dimensions are in inches.
2. Material most commonly used, forged steel SA 105. Available also in stainless steel, alloy steel, and nonferrous metal.
3. The 1/4 in. raised face is included in thickness J but is included in length M.
4. The length of bolts does not include the height of crown.
5. Bolt holes are 1/8 in. larger than bolt diameters.
6. Dimensions, M (length of welding necks), are based on data of major manufacturers. Long welding necks with necks longer than listed are available on special order.

See Facing Page for Dimension J.

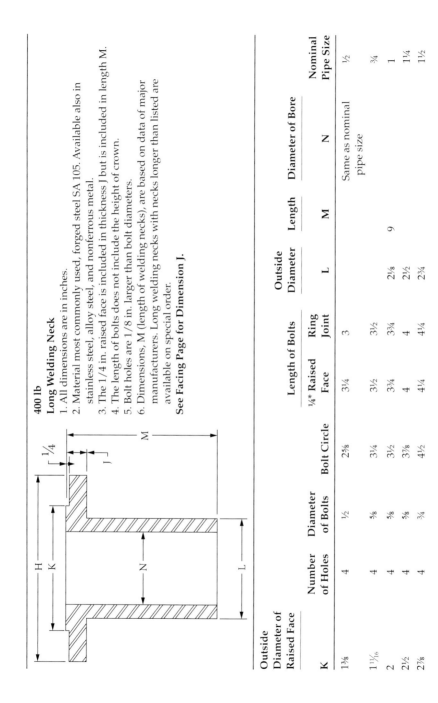

Outside Diameter of Raised Face K	Number of Holes	Diameter of Bolts	Bolt Circle	Length of Bolts ¼* Raised Face	Length of Bolts Ring Joint	Outside Diameter L	Length M	Diameter of Bore N	Nominal Pipe Size
1⅜	4	½	2⅝	3¼	3				½
1¹¹⁄₁₆	4	⅝	3¼	3½	3½			Same as nominal pipe size	¾
2	4	⅝	3½	3¾	3¾	2⅛			1
2½	4	⅝	3⅞	4	4	2½	9		1¼
2⅞	4	¾	4½	4¼	4¼	2¾			1½

3⅝	8	⅝	5	4¼	4½	3 5/16		2
4⅛	8	¾	5⅞	4¾	5	3 15/16		2½
5	8	¾	6⅝	5	5¼	4⅝		3
5½	8	⅞	7¼	5½	5¾	5¼		3½
6 3/16	8	⅞	7⅞	5½	5¾	5¾	12	4
7 5/16	8	⅞	9¼	5¾	6	7		5
8½	12	⅞	10⅝	6	6¼	8⅛		6
10⅝	12	1	13	6¾	7	10¼		8
12¾	16	1⅛	15¼	7½	7¾	12⅝		10
15	16	1¼	17¾	8	8¼	14¾		12
16¼	20	1¼	20¼	8¼	8½	16¾	10–14	14
18½	20	1⅜	22½	8¾	9	19		16
21	24	1⅜	24¾	9	9¼	21		18
23	24	1½	27	9¾	10	23⅛		20
25¼	24	1⅝	29¼	10	10½			22
27¼	24	1¾	32	10¾	11¼	27⅝		24
29½	28	1¾	34½	11½	12			26
31½	28	1⅞	37	12¼	12¾			28
33¾	28	2	39¼	13	13½			30

600 lb Flanges
Standard ANSI B16.5

1. All dimensions are in inches.
2. Material most commonly used, forged steel SA 105. Available also in stainless steel, alloy steel, and nonferrous metal.
3. The 1/4 in. raised face is not included in dimensions C, D, and J.
4. The lengths of stud bolts do not include the height of crown.
5. Bolt holes are 1/8 in. larger than bolt diameters.
6. Flanges bored to dimensions shown unless otherwise specified.
7. Flanges for pipe sizes 22, 26, 28, and 30 are not covered by ANSI B16.5

See Facing Page for Dimension K and Data on Bolting.

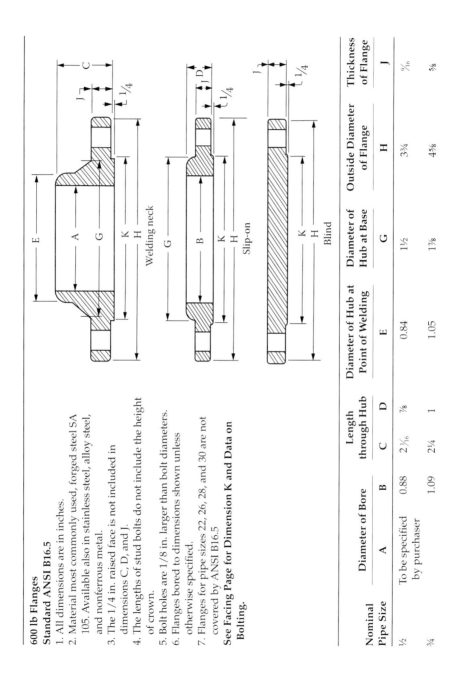

Welding neck

Slip-on

Blind

Nominal Pipe Size	Diameter of Bore		Length through Hub		Diameter of Hub at Point of Welding	Diameter of Hub at Base	Outside Diameter of Flange	Thickness of Flange
	A	B	C	D	E	G	H	J
½	To be specified by purchaser	0.88	2 $\frac{1}{16}$	⅞	0.84	1½	3¾	$\frac{9}{16}$
¾		1.09	2¼	1	1.05	1⅞	4⅝	⅝

1	1.36	2⁷⁄₁₆	1¹⁄₁₆	1.32	2⅛	4⅞	¹¹⁄₁₆
1¼	1.70	2⅝	1⅛	1.66	2½	5¼	¹³⁄₁₆
1½	1.95	2¾	1¼	1.90	2¾	6⅛	⅞
2	2.44	2⅞	1⁷⁄₁₆	2.38	3⁵⁄₁₆	6½	1
2½	2.94	3⅛	1⅝	2.88	3¹⁵⁄₁₆	7½	1⅛
3	3.57	3¼	1¹³⁄₁₆	3.50	4⅝	8¼	1¼
3½	4.07	3⅜	1¹⁵⁄₁₆	4.00	5¼	9	1⅜
4	4.57	4	2⅛	4.50	6	10¾	1½
5	5.66	4½	2⅜	5.56	7⁷⁄₁₆	13	1¾
6	6.72	4⅝	2⅝	6.63	8¾	14	1⅞
8	8.72	5¼	3	8.63	10¾	16½	2³⁄₁₆
10	10.88	6	3⅜	10.75	13½	20	2½
12	12.88	6⅛	3⅝	12.75	15¾	22	2⅝
14	14.14	6½	3¹¹⁄₁₆	14.00	17	23¾	2¾
16	16.16	7	4³⁄₁₆	16.00	19½	27	3
18	18.18	7¼	4⅝	18.00	21½	29¼	3¼
20	20.20	7½	5	20.00	24	32	3½
22	22.22	7¾	5¼	22.00	26¼	34¼	3¾
24	24.25	8	5½	24.00	28¼	37	4
26	26.25	8¾	8¾	26⁷⁄₁₆	29⁷⁄₁₆	40	4¼
28	28.25	9¼	9¼	28½	31⅝	42¼	4⅜
30	30.25	9¾	9¾	30½	33¹⁵⁄₁₆	44½	4½

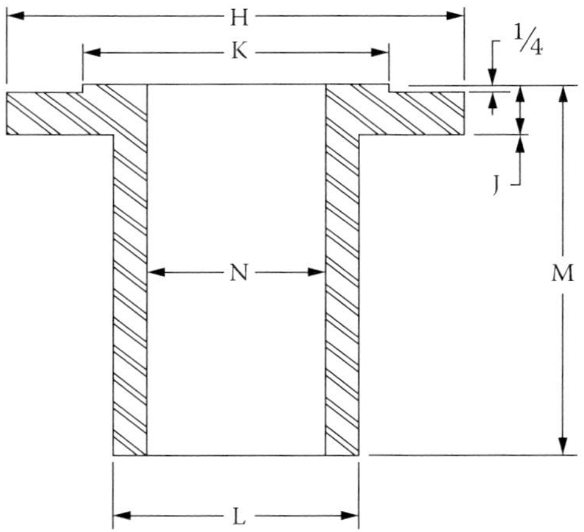

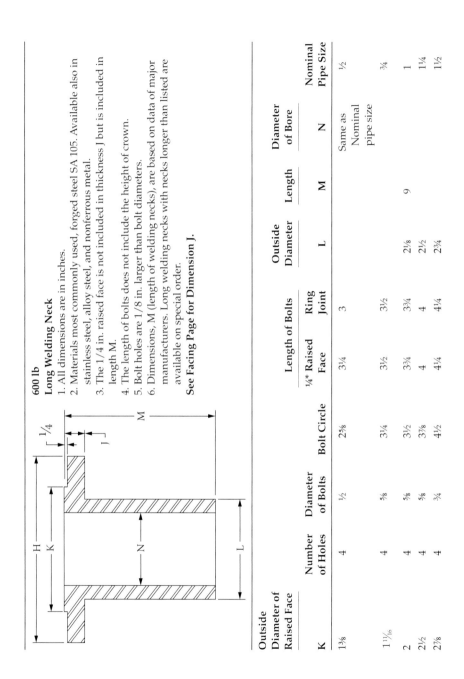

600 lb

Long Welding Neck

1. All dimensions are in inches.
2. Materials most commonly used, forged steel SA 105. Available also in stainless steel, alloy steel, and nonferrous metal.
3. The 1/4 in. raised face is not included in thickness J but is included in length M.
4. The length of bolts does not include the height of crown.
5. Bolt holes are 1/8 in. larger than bolt diameters.
6. Dimensions, M (length of welding necks), are based on data of major manufacturers. Long welding necks with necks longer than listed are available on special order.

See Facing Page for Dimension J.

Outside Diameter of Raised Face				Length of Bolts		Outside Diameter	Length	Diameter of Bore	
K	Number of Holes	Diameter of Bolts	Bolt Circle	¼* Raised Face	Ring Joint	L	M	N	Nominal Pipe Size
1⅜	4	½	2⅝	3¼	3			Same as Nominal pipe size	½
1 11/16	4	⅝	3¼	3½	3½				¾
2	4	⅝	3½	3¾	3¾	2⅛	9		1
2½	4	⅝	3⅞	4	4	2½			1¼
2⅞	4	¾	4½	4¼	4¼	2¾			1½

3⅝	8	⅝	5	4¼	4½	3¹⁵⁄₁₆		2
4⅛	8	¾	5⅞	4¾	5	3¹⁵⁄₁₆		2½
5	8	¾	6⅝	5	5¼	4⅝		3
5½	8	⅞	7¼	5½	5¾	5¼		3½
6³⁄₁₆	8	⅞	8½	5¾	6	6	12	4
7⁵⁄₁₆	8	1	10½	6½	6¾	7½		5
8½	12	1	11½	6¾	7	8¾		6
10⅝	12	1⅛	13¾	7¾	7¾	10¾		8
12¾	16	1¼	17	8½	8¾	13½		10
15	20	1¼	19¼	8¾	9	15¾		12
16¼	20	1⅜	20¾	9¼	9½	17	12–20	14
18½	20	1½	23¾	10	10¼	19½		16
21	20	1⅝	25¾	10¾	11	21½		18
23	24	1⅝	28½	11½	11¾	24		20
25¼	24	1¾	30⅝	12	12½			22
27¼	24	1⅞	33	13	13¼	28¼		24
29½	28	1⅞	36	13¼	13¾			26
31½	28	2	38	13¾	14¼			28
33¾	28	2	40¼	14	14½			30

900 lb Flanges
Standard ANSI B16.5

1. All dimensions are in inches.
2. Material most commonly used, forged steel SA 105. Available also in stainless steel, alloy steel, and nonferrous metal.
3. The 1/4 in. raised face is included in dimensions C, D, and J.
4. The lengths of stud bolts do not include the height of crown.
5. Bolt holes are 1/8 in. larger than bolt diameters.
6. Flanges bored to dimensions shown unless otherwise specified.
7. Flanges for pipe sizes 26, 28, and 30 are not covered by ANSI B16.5

See Facing Page for Dimension K and Data on Bolting.

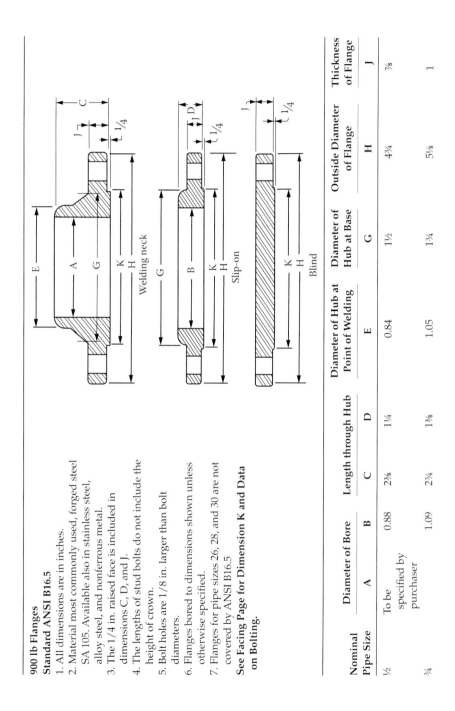

Nominal Pipe Size	Diameter of Bore		Length through Hub		Diameter of Hub at Point of Welding	Diameter of Hub at Base	Outside Diameter of Flange	Thickness of Flange
	A	B	C	D	E	G	H	J
½	To be specified by purchaser	0.88	2⅜	1¼	0.84	1½	4¾	⅞
¾		1.09	2¾	1⅜	1.05	1¾	5⅛	1

1	1.36	2⅞	1⅝	1.32	2 1/16	5⅞	1⅛
1¼	1.70	2⅞	1⅝	1.66	2½	6¼	1⅛
1½	1.95	3¼	1¾	1.90	2¾	7	1¼
2	2.44	4	2¼	2.38	4⅛	8½	1½
2½	2.94	4⅛	2½	2.88	4⅞	9⅜	1⅝
3	3.57	4	2⅛	3.50	5	9½	1½
4	4.57	4½	2¾	4.50	6¼	11½	1¾
5	5.66	5	3⅛	5.56	7½	13¾	2
6	6.72	5½	3⅜	6.63	9¼	15	2 3/16
8	8.72	6⅜	4	8.63	11¾	18½	2½
10	10.88	7¼	4¼	10.75	14½	21½	2¾
12	12.88	7⅞	4⅝	12.75	16½	24	3⅛
14	14.14	8⅜	5⅛	14.00	17¾	25¼	3⅜
16	16.16	8½	5¼	16.00	20	27¾	3½
18	18.18	9	6	18.00	22¼	31	4
20	20.20	9¾	6¼	20.00	24½	33¾	4¼
24	24.25	11½	8	24.00	29½	41	5½
26	26.25	11¼	11¼	26⅝	30½	42¾	5½
28	28.25	11¾	11¾	28 11/16	32¾	46	5⅝
30	30.25	12¼	12¼	30¾	35	48½	5⅞

900 lb
Long Welding Neck

1. All dimensions are in inches.
2. Material most commonly used, forged steel SA 105. Available also in stainless steel, alloy steel, and nonferrous metal.
3. The 1/4 in. raised face is not included in thickness J but is included in length M.
4. The length of bolts does not include the height of crown.
5. Bolt holes are 1/8 in. larger than bolt diameters.
6. Dimensions, M (length of welding necks), are based on data of major manufacturers. Long welding necks with necks longer than listed are available on special order.

See Facing Page for Dimension J.

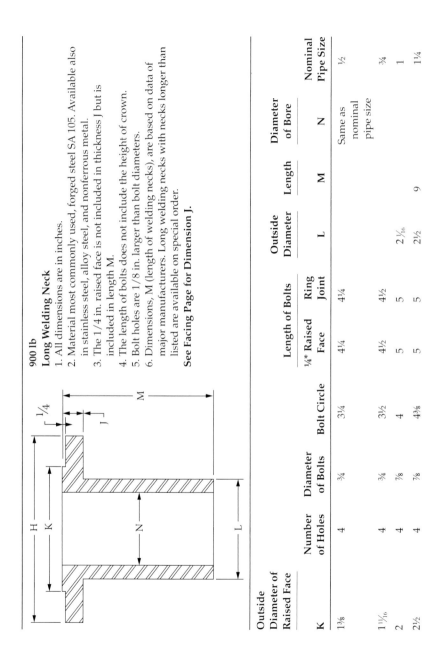

Outside Diameter of Raised Face K	Number of Holes	Diameter of Bolts	Bolt Circle	Length of Bolts ¼* Raised Face	Length of Bolts Ring Joint	Outside Diameter L	Length M	Diameter of Bore N	Nominal Pipe Size
1⅜	4	¾	3¼	4¼	4¼			Same as nominal pipe size	½
1¹¹⁄₁₆	4	¾	3½	4½	4½				¾
2	4	⅞	4	5	5	2¹⁄₁₆			1
2½	4	⅞	4⅜	5	5	2½	9		1¼

2⅞	4	1	4⅞	5½	5½	2¾		1½
3⅝	8	⅞	6½	5¾	5¾	4⅛		2
4⅛	8	1	7½	6¼	6¼	4⅞	12	2½
5	8	⅞	7½	5¾	6	5		3
6 3/16	8	1⅛	9¼	6¾	7	6¼		4
7 5/16	8	1¼	11	7½	7¾	7½		5
8½	12	1⅛	12½	7¾	7¼	9¼		6
10⅝	12	1⅜	15½	8¾	9	11¾		8
12¾	16	1⅜	18½	9¼	9½	14½	12–20	10
15	20	1⅜	21	10	10¼	16½		12
16¼	20	1½	22	10¾	11¼	17¾		14
18½	20	1⅝	24¼	11¼	11¾	20		16
21	20	1⅞	27	12¾	13½	22¼		18
23	20	2	29½	13½	14¼	24½		20
27¼	20	2½	35½	17¼	17¾	29½		24
29½	20	2¾	37½	17½	18¾			26
31½	20	3	40¼	18¼	19½			28
33¾	20	3	42¾	18¾	20			30

1500 lb Flanges
Standard ANSI B16.5

1. All dimensions are in inches.
2. Material most commonly used, forged steel SA 105. Available also in stainless steel, alloy steel, and nonferrous metal.
3. The 1/4 in. raised face is not included in dimensions C, D, and J.
4. The lengths of stud bolts do not include the height of crown.
5. Bolt holes are 1/8 in. larger than bolt diameters.
6. Flanges bored to dimensions shown unless otherwise specified.

See Facing Page for Dimension K and Data on Bolting.

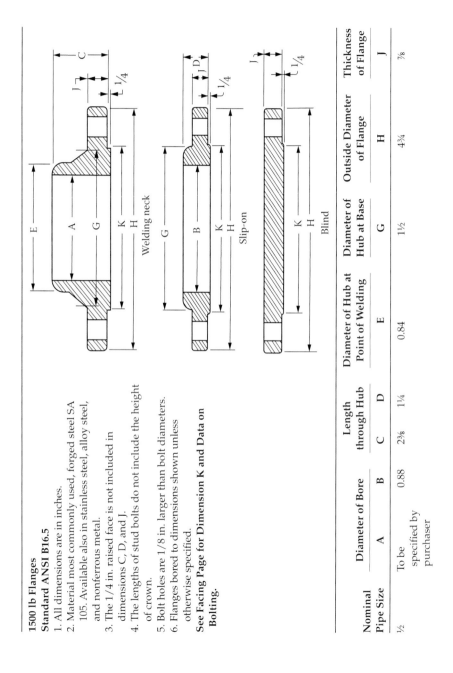

Welding neck

Slip-on

Blind

| Nominal Pipe Size | Diameter of Bore | | Length through Hub | | Diameter of Hub at Point of Welding | Diameter of Hub at Base | Outside Diameter of Flange | Thickness of Flange |
	A	B	C	D	E	G	H	J
½	To be specified by purchaser	0.88	2⅜	1¼	0.84	1½	4¾	⅞

¾	1.09	2¾	1⅜	1.05	1¾	5⅛	1
1	1.36	2⅞	1⅝	1.32	2 1/16	5⅞	1⅛
1¼	1.70	2⅞	1⅝	1.66	2½	6¼	1⅛
1½	1.95	3¼	1¾	1.90	2¾	7	1¼
2	2.44	4	2¼	2.38	4⅛	8½	1½
2½	2.94	4⅛	2½	2.88	4⅞	9⅝	1⅝
3	3.57	4⅝	2⅞	3.50	5¼	10½	1⅞
4	4.57	4⅞	3 9/16	4.50	6⅜	12¼	2⅛
5	5.66	6⅛	4⅛	5.56	7¾	14¾	2⅞
6	6.72	6¾	4 11/16	6.63	9	15½	3¼
8	8.72	8⅜	5⅝	8.63	11½	19	3⅝
10	10.88	10	6¼	10.75	14½	23	4¼
12	12.88	11⅛	7⅛	12.75	17¾	26½	4⅞
14	—	11¾	—	14.00	19½	29½	5¼
16	—	12¼	—	16.00	21¾	32½	5¾
18	—	12⅞	—	18.00	23½	36	6⅜
20	—	14	—	20.00	25¼	38¾	7
24	—	16	—	24.00	30	46	8

1500 lb

Long Welding Neck

1. All dimensions are in inches.
2. Material most commonly used, forged steel SA 105. Available also in stainless steel, alloy steel, and nonferrous metal.
3. The 1/4 in. raised face is not included in thickness J but is included in length M.
4. The length of bolts does not include the height of crown.
5. Bolt holes are 1/8 in larger than bolt diameters.
6. Dimensions, M (length of welding necks), are based on data of major manufacturers. Long welding necks with necks longer than listed are available on special order.

See Facing Page for Dimension J.

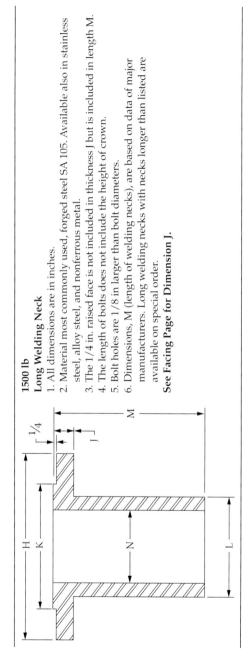

Outside Diameter of Raised Face K	Number of Holes	Diameter of Bolts	Bolt Circle	Length of Bolts 1/4* Raised Face	Length of Bolts Ring Joint	Outside Diameter L	Length M	Diameter of Bore N	Nominal Pipe Size
1 3/8	4	3/4	3 1/4	4 1/4	4 1/4			Same as nominal pipe size	1/2
1 11/16	4	3/4	3 1/2	4 1/2	4 1/2				3/4
2	4	7/8	4	5	5	2 1/16	9		1
2 1/2	4	7/8	4 3/8	5	5	2 1/2			1 1/4
2 7/8	4	1	4 7/8	5 1/2	5 1/2	2 3/4			1 1/2
3 5/8	8	7/8	6 1/2	5 3/4	5 3/4	4 1/8			2

4⅛	8	1	7½	6¼	6¼	4⅞	12	2½
5	8	1⅛	8	7	7	5¼		3
6 3⁄16	8	1¼	9½	7¾	7¾	6⅜		4
7 5⁄16	8	1½	11½	9¾	9¾	7¾		5
8½	12	1⅜	12½	10¼	10½	9		6
10⅝	12	1⅝	15½	11½	12	11½		8
12¾	12	1⅞	19	13¼	13¾	14½	12–20	10
15	16	2	22½	14¾	15½	17¾		12
16¼	16	2¼	25	16	17	19½		14
18½	16	2½	27¾	17½	18½	21¾		16
21	16	2¾	30½	19½	20½	23½		18
23	16	3	32¾	21½	22½	25¼		20
27¼	16	3½	39	24½	25¾	30		24

2500 lb Flanges
Standard ANSI B16.5

1. All dimensions are in inches.
2. Material most commonly used, forged steel SA 105. Available also in stainless steel, alloy steel, and nonferrous metal.
3. The 1/4 in. raised face is not included in dimensions C, D, and J.
4. The lengths of stud bolts do not include the height of crown.
5. Bolt holes are 1/8 in. larger than bolt diameters.
6. Flanges bored to dimensions shown unless otherwise specified.
See Facing Page for Dimension K and Data on Bolting.

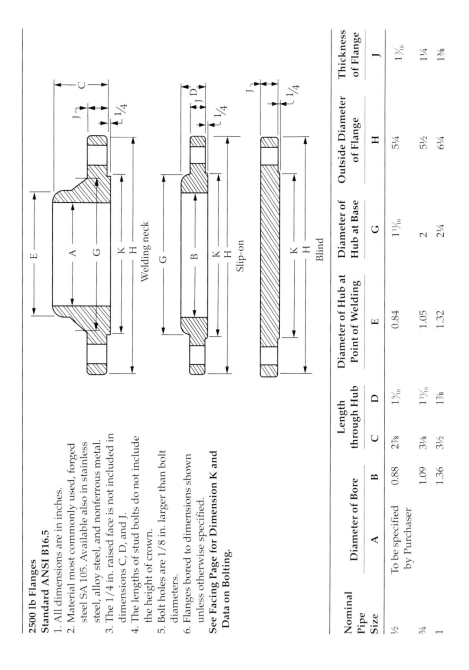

Welding neck

Slip-on

Blind

Nominal Pipe Size	Diameter of Bore		Length through Hub		Diameter of Hub at Point of Welding	Diameter of Hub at Base	Outside Diameter of Flange	Thickness of Flange
	A	B	C	D	E	G	H	J
½	To be specified by Purchaser	0.88	2⅞	1⁹⁄₁₆	0.84	1¹¹⁄₁₆	5¼	1³⁄₁₆
¾		1.09	3⅛	1¹¹⁄₁₆	1.05	2	5½	1¼
1		1.36	3½	1⅞	1.32	2¼	6¼	1⅜

1¼	1.70	3¾	2 1/16	1.66	2⅞	7¼	1½
1½	1.95	4⅜	2⅜	1.90	3⅜	8	1¾
2	2.44	5	2¾	2.38	3¾	9¼	2
2½	2.94	5⅝	3⅛	2.88	4½	10½	2¼
3	3.57	6⅝	3⅝	3.50	5¼	12	2⅝
4	4.57	7½	4¼	4.50	6½	14	3
5	5.66	9	5⅛	5.56	8	16½	3⅝
6	6.72	10¾	6	6.63	9¼	19	4¼
8	8.72	12½	7	8.63	12	21¾	5
10	10.88	16½	9	10.75	14¾	26½	6½
12	12.88	18¼	10	12.75	17⅜	30	7¼

2500 lb
Long Welding Neck

1. All dimensions are in inches.
2. Material most commonly used, forged steel SA 105. Available also in stainless steel, alloy steel, and nonferrous metal.
3. The 1/4 in. raised face is not included in thickness J but is included in length M.
4. The length of bolts does not include the height of crown.
5. Bolt holes are 1/8 in. larger than bolt diameters.
6. Dimensions, M (length of welding necks), are based on data of major manufacturers. Long welding necks with necks longer than listed are available on special order.

See Facing Page For Dimension J.

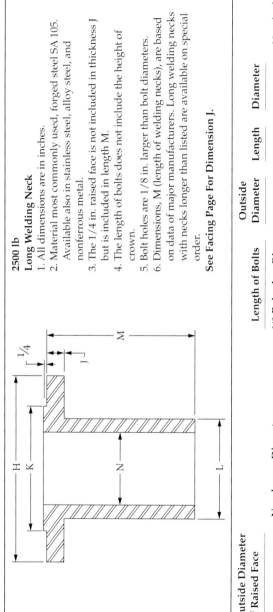

Outside Diameter of Raised Face K	Number of Holes	Diameter of Bolts	Bolt Circle	Length of Bolts 1/4* Raised Face	Ring Joint	Outside Diameter L	Length M	Diameter N	Nominal Pipe Size
1⅜	4	¾	3½	5¼	5¼		9	Same as nominal pipe size	½
1¹¹/₁₆	4	¾	3¾	5¼	5¼				¾
2	4	⅞	4¼	5¾	5¾	2¼			1
2½	4	1	5⅛	6¼	6½	2⅞	12		1¼
2⅞	4	1⅛	5¾	7	7¼	3⅛			1½
3⅝	8	1	6¾	7¼	7½	3¾			2

4⅛	8	1⅛	7¾	8	8¼	4½		2½
5	8	1¼	9	9	9¼	5¼		3
6 3/16	8	1½	10¾	10¼	10¾	6½		4
7 5/16	8	1¾	12¾	12	12¾	8		5
8½	8	2	14½	13¼	14½	9¼		6
10⅝	12	2	17¼	15¼	16	12		8
12¾	12	2½	21¼	19½	20½	14¾	12–20	10
15	12	2¾	24⅜	21½	22½	17⅜		12

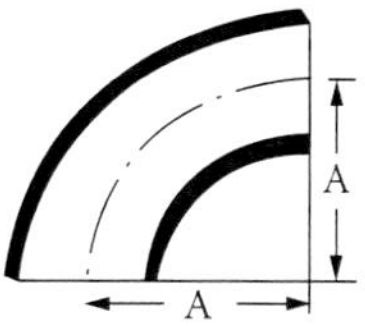

90° Long radius elbow

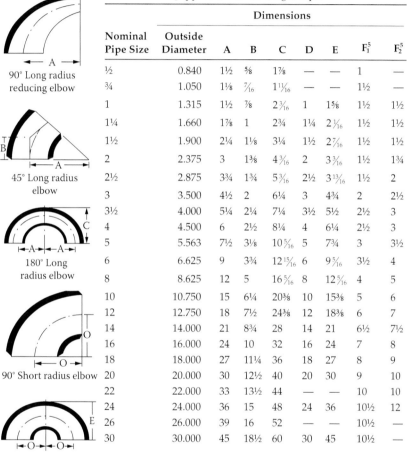

90° Long radius
reducing elbow

45° Long radius
elbow

180° Long
radius elbow

90° Short radius elbow

180° Short radius return

Welding Fittings
ANSI B16.9
1. All dimensions are in inches.
2. Welding fitting material conforms to SA 234 grade WPB.
3. Sizes 22, 26, and 30 in. are not covered by ANSI B16.9.
4. For wall thickness, see p. 322.
5. Dimension F_1 applies to standard and X-STG caps.
   Dimension F_2 applies to heavier weight caps.

		Dimensions						
Nominal Pipe Size	Outside Diameter	A	B	C	D	E	F_1^5	F_2^5
½	0.840	1½	⅝	1⅞	—	—	1	—
¾	1.050	1⅛	⁷⁄₁₆	1¹¹⁄₁₆	—	—	1½	—
1	1.315	1½	⅞	2³⁄₁₆	1	1⅝	1½	1½
1¼	1.660	1⅞	1	2¾	1¼	2¹⁄₁₆	1½	1½
1½	1.900	2¼	1⅛	3¼	1½	2⁷⁄₁₆	1½	1½
2	2.375	3	1⅜	4³⁄₁₆	2	3³⁄₁₆	1½	1¾
2½	2.875	3¾	1¾	5³⁄₁₆	2½	3¹³⁄₁₆	1½	2
3	3.500	4½	2	6¼	3	4¾	2	2½
3½	4.000	5¼	2¼	7¼	3½	5½	2½	3
4	4.500	6	2½	8¼	4	6¼	2½	3
5	5.563	7½	3⅛	10⁵⁄₁₆	5	7¾	3	3½
6	6.625	9	3¾	12¹⁵⁄₁₆	6	9⁵⁄₁₆	3½	4
8	8.625	12	5	16⁵⁄₁₆	8	12⁵⁄₁₆	4	5
10	10.750	15	6¼	20⅜	10	15⅜	5	6
12	12.750	18	7½	24⅜	12	18⅜	6	7
14	14.000	21	8¾	28	14	21	6½	7½
16	16.000	24	10	32	16	24	7	8
18	18.000	27	11¼	36	18	27	8	9
20	20.000	30	12½	40	20	30	9	10
22	22.000	33	13½	44	—	—	10	10
24	24.000	36	15	48	24	36	10½	12
26	26.000	39	16	52	—	—	10½	—
30	30.000	45	18½	60	30	45	10½	—

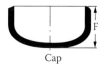

Cap

Welding Fittings
ANSI B16.9
1. All dimensions are in inches.
2. Welding fitting material conforms to SA 234 grade WPB.
3. Sizes 22, 26, and 30 in. are not covered by ANSI B16.9.
4. For wall thicknesses, see p. 322.

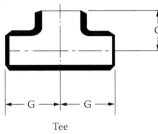

Tee

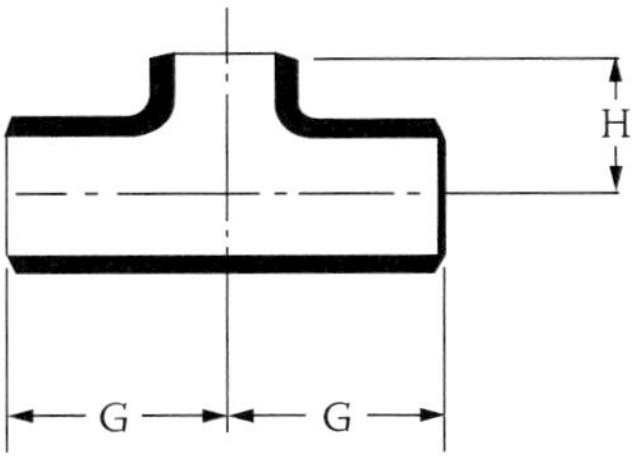

Reducing tee

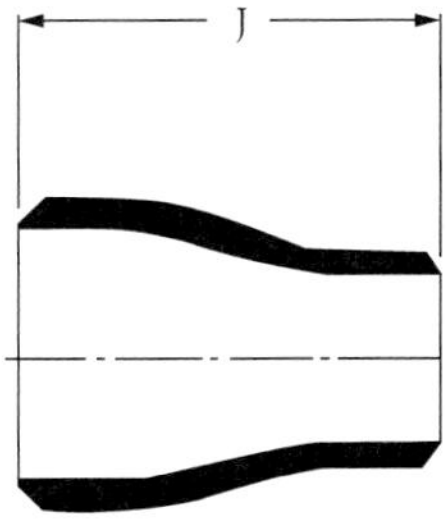

Concentric reducer

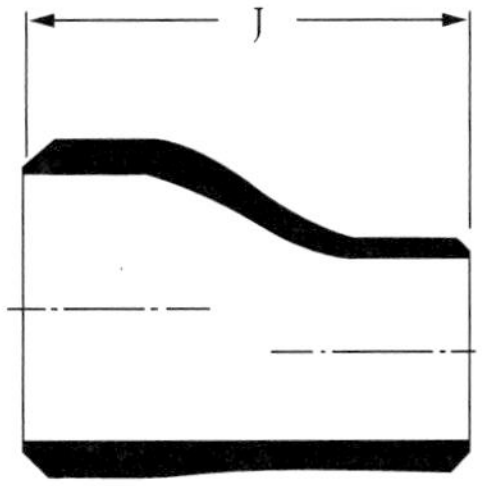

Eccentric reducer

Nominal Pipe Size	Dimensions				
	Outlet	Outside Diameter	G	H	J
½	½	0.840	1	1	—
	⅜	0.675	1	1	—
¾	¾	1.050	1⅛	1⅛	—
	½	0.840	1⅛	1⅛	1½
1	1	1.315	1½	1½	—
	¾	1.050	1½	1½	2
	½	0.840	1½	1½	2
1¼	1¼	1.660	1⅞	1⅞	—
	1	1.315	1⅞	1⅞	2
	¾	1.050	1⅞	1⅞	2
	½	0.840	1⅞	1⅞	2
1½	1½	1.900	2¼	2¼	—
	1¼	1.660	2¼	2¼	2½
	1	1.315	2¼	2¼	2½
	¾	1.050	2¼	2¼	2½
	½	0.840	2¼	2¼	2½
2	2	2.375	2½	2½	—
	1½	1.900	2½	2⅜	3
	1¼	1.660	2½	2⅛	3
	1	1.315	2½	2	3
	¾	1.050	2½	1¾	3
2½	2½	2.875	3	3	—
	2	2.375	3	2¾	3½
	1½	1.900	3	2⅝	3½
	1¼	1.660	3	2½	3½
	1	1.315	3	2¼	3½
3	3	3.500	3⅜	3⅜	—
	2½	2.875	3⅜	3¼	3½
	2	2.375	3⅜	3	3½
	1½	1.900	3⅜	2⅞	3½
	1¼	1.660	3⅜	2¾	3½
3½	3½	4.000	3¾	3¾	—
	3	3.500	3¾	3⅝	4
	2½	2.875	3¾	3½	4
	2	2.375	3¾	3¼	4
	1½	1.900	3¾	3⅛	4

(continued)

(continued)

Nominal Pipe Size		Dimensions			
	Outlet	Outside Diameter	G	H	J
4	4	4.500	4⅛	4⅛	—
	3½	4.000	4⅛	4	4
	3	3.500	4⅛	3⅞	4
	2½	2.875	4⅛	3¾	4
	2	2.375	4⅛	3½	4
	1½	1.900	4⅛	3⅜	4

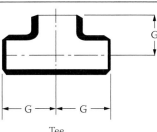

Tee

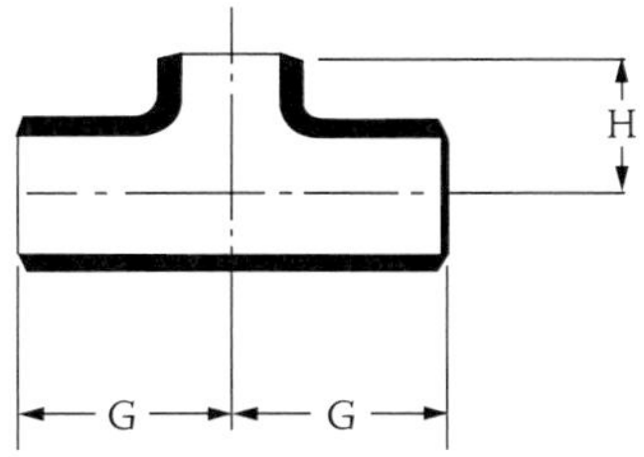

Reducing tee

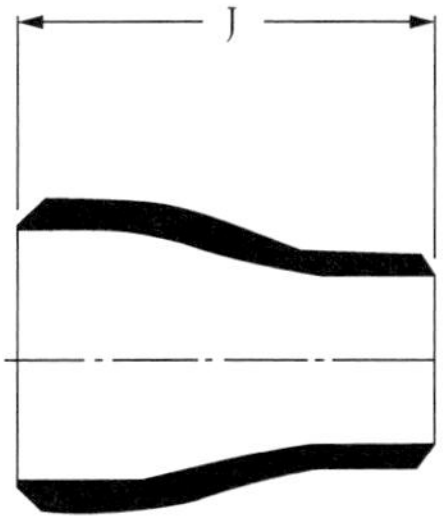

Concentric reducer

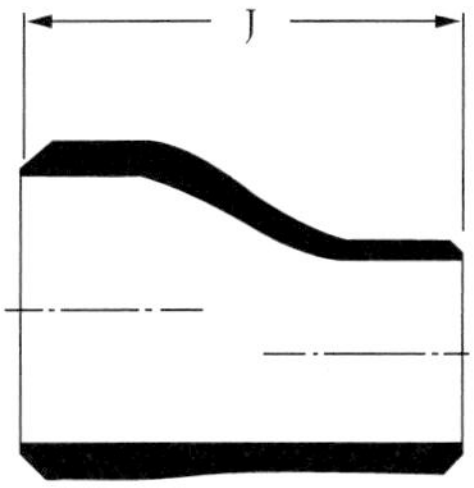

Eccentric reducer

Welding Fittings
ANSI B16.9
1. All dimensions are in inches.
2. Welding fitting material conforms to SA 234 grade WPB.
3. Sizes 22, 26, and 30 in. are not covered by ANSI B16.9.
4. For wall thicknesses, see p. 322.

Nominal Pipe Size	Outlet	Outside Diameter	G	H	J
5	5	5.563	4⅞	4⅞	—
	4	4.500	4⅞	4⅝	5
	3½	4.000	4⅞	4½	5
	3	3.500	4⅞	4⅜	5
	2½	2.875	4⅞	4¼	5
	2	2.375	4⅞	4⅛	5
6	6	6.625	5⅝	5⅝	—
	5	5.563	5⅝	5⅜	5½
	4	4.500	5⅝	5⅛	5½
	3½	4.000	5⅝	5	5½
	3	3.500	5⅝	4⅞	5½
	2½	2.875	5⅝	4¾	5½
8	8	8.625	7	7	—
	6	6.625	7	6⅝	6
	5	5.536	7	6⅜	6
	4	4.500	7	6⅛	6
	3½	4.000	7	6	6
10	10	10.750	8½	8½	—
	8	8.625	8½	8	7
	6	6.625	8½	7⅝	7
	5	5.563	8½	7½	7
	4	4.500	8½	7¼	7
12	12	12.750	10	10	—
	10	10.750	10	9½	8
	8	8.625	10	9	8
	6	6.625	10	8⅜	8
	5	5.563	10	8½	8
14	14	14.00	11	11	—
	12	12.750	11	10⅝	13
	10	10.750	11	10⅛	13
	8	8.625	11	9¾	13
	6	6.625	11	9⅜	13

(continued)

(continued)

Nominal Pipe Size	Dimensions				
	Outlet	Outside Diameter	G	H	J
16	16	16.000	12	12	—
	14	14.000	12	12	14
	12	12.750	12	11⅝	14
	10	10.750	12	11⅛	14
	8	8.625	12	10¾	14
	6	6.625	12	10⅛	14
18	18	18.000	13½	13½	—
	16	16.000	13½	13	15
	14	14.000	13½	13	15

Welding Fittings
ANSI B16.9
1. All dimensions are in inches.
2. Welding fitting material conforms to SA 234 grade WPB.
3. Sizes 22, 26, and 30 in. are not covered by ANSI B16.9.
4. For wall thicknesses, see p. 322.

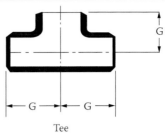

Tee

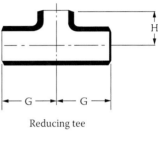

Reducing tee

Nominal Pipe Size	Dimensions				
	Outlet	Outside Diameter	G	H	J
18	12	12.750	13½	12⅝	15
	10	10.750	13½	12⅛	15
	8	8.625	13½	11¾	15
20	20	20.000	15	15	—
	18	18.000	15	14½	20
	16	16.000	15	14	20
	14	14.000	15	14	20
	12	12.750	15	13⅝	20
	10	10.750	15	13⅛	20
	8	8.625	15	12¾	20
22	22	22.000	16½	16½	—
	20	20.000	16½	16	20
	18	18.000	16½	15½	20
	16	16.000	16½	15	20
	14	14.000	16½	15	20
	12	12.750	16½	14⅝	—
	10	10.750	16½	14⅛	—
24	24	24.000	17	17	—
	22	22.000	17	17	20
	20	20.000	17	17	20
	18	18.000	17	16½	20
	16	16.000	17	16	20
	14	14.000	17	16	20
	12	12.750	17	15⅝	20
	10	10.750	17	15⅛	20
30	30	30.000	22	22	—
	24	24.000	22	21	24
	22	22.000	22	20½	24
	20	20.000	22	20	24
	18	18.000	22	19½	—
	16	16.000	22	19	—

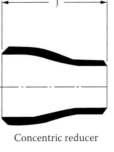

Concentric reducer

Eccentric reducer

Face-to-Face Dimensions of
Flanged Steel Gate Valves
(Wedge and Double Disc)

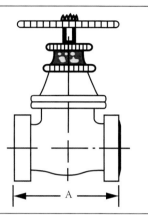

	Raised Face							
	Pressure (lb per sq. in.)					Pressure (lb per sq. in.)		
Nominal Size (in.)	150	300	400	600	Nominal Size (in.)	900	1500	2500
	Dimension A (in.)					Dimension A (in.)		
1	—	—	8½	8½	1	10	10	12⅛
1¼	—	—	9	9	1¼	11	11	13¾
1½	—	7½	9½	9½	1½	12	12	15⅛
2	7	8½	11½	11½	2	14½	14½	17¾
2½	7½	9½	13	13	2½	16½	16½	20
3	8	11⅛	14	14	3	15	18½	22¾
3½	8½	11⅞	—	—	4	18	21½	26½
4	9	12	16	17	5	22	26½	31¼
5	10	15	18	20	6	24	27¾	36
6	10½	15⅞	19½	22	8	29	32¾	40¼
8	11½	16½	23½	26	10	33	39	50
10	13	18	26½	31	12	38	44½	56
12	14	19¾	30	33	14	40½	49½	—
14 OD	15	30	32½	35	16	44½	54½	—
16 OD	16	33	35½	39	18	48	60½	—
18 OD	17	36	38½	43	20	52	65½	—
20 OD	18	39	41½	47	24	61	76½	—
24 OD	20	45	48½	55				

(continued)

	Ring-Type Joint							
	Pressure (lb per sq. in.)					Pressure (lb per sq. in.)		
Nominal Size (in.)	150	300	400	600	Nominal Size (in.)	900	1500	2500
	Dimension A (in.)					Dimension A (in.)		
1	5½	—	8½	8½	1	10	10	12⅛
1¼	6	—	9	9	1¼	11	11	13⅞
1½	7	8	9½	9½	1½	12	12	15¼
2	7½	9⅛	11⅝	11⅝	2	14⅝	14⅝	17⅞
2½	8	10⅛	13⅛	13⅛	2½	16⅝	16⅝	20¼
3	8½	11¾	14⅛	14⅛	3	15⅛	18⅝	23
4	9½	12⅝	16⅞	17⅛	4	18⅛	21⅝	26⅞
5	10½	15⅝	18⅞	20⅛	5	22⅛	26⅝	31¾
6	11	16½	19⅝	22⅛	6	24⅛	28	36½
8	12	17⅛	23⅝	26⅛	8	29⅛	33⅛	40⅞
10	13½	18⅝	26⅝	31⅛	10	33⅛	39⅜	50⅞
12	14½	20⅜	30⅛	33⅛	12	38⅛	45⅛	56⅞
14	15½	30⅝	32⅝	35⅛	14	40⅞	50¼	—
16	16½	33⅝	35⅝	39⅛	16	44⅞	55⅜	—
18	17½	36⅝	38⅝	43⅛	18	48½	61⅜	—
20	18½	39¾	41¾	47¼	20	52½	66⅜	—
24	20½	45⅞	48⅞	55¾	24	61¾	77⅝	—

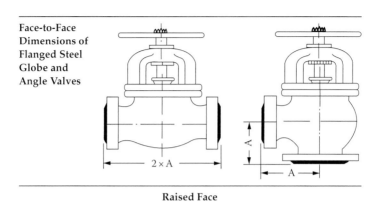

Face-to-Face
Dimensions of
Flanged Steel
Globe and
Angle Valves

Raised Face

	Class (lb)					Pressure (lb per sq. in.)		
	150	300	400	600		900	1500	2500
Nominal Size (in.)	Dimension 2 × A (in.)				Nominal Size (in.)	Dimension 2 × A (in.)		
½	—	—	—	—	½	—	—	10⅜
¾	—	—	7½	7½	¾	9	9	10¾
1	—	—	8½	8½	1	10	10	12⅛
1¼	—	—	9	9	1¼	11	11	13¾
1½	—	—	9½	9½	1½	12	12	15⅛
2	8	10½	11½	11½	2	14½	14½	17¾
2½	8½	11½	13	13	2½	16½	16½	20
3	9½	12½	14	14	3	15	18½	22¾
3½	10½	13¼	—	—	4	18	21½	26½
4	11½	14	16	17	5	22	26½	31¼
5	14	15¾	18	20	6	24	27¾	36
6	16	17½	19½	22	8	29	32¾	40¼
8	19½	22	23½	26	10	33	39	50
					12	38	44½	56
					14	40½	49½	—

(continued)

	Ring-Type Joint							
	Pressure (lb per sq. in.)					Pressure (lb per sq. in.)		
	150	300	400	600		900	1500	2500
Nominal Size (in.)	Dimension 2 × A (in.)				Nominal Size (in.)	Dimension 2 × A (in.)		
½	—	$6\frac{7}{16}$	$6\frac{7}{16}$	$6\frac{7}{16}$	½	—	—	$10\frac{3}{8}$
¾	—	7½	7½	7½	¾	9	9	10¾
1	—	8½	8½	8½	1	10	10	$12\frac{1}{8}$
1¼	—	9	9	9	1¼	11	11	$13\frac{7}{8}$
1½	7	9½	9½	9½	1½	12	12	15¼
2	8½	$11\frac{1}{8}$	$11\frac{5}{8}$	$11\frac{5}{8}$	2	$14\frac{5}{8}$	$14\frac{5}{8}$	$17\frac{7}{8}$
2½	9	$12\frac{1}{8}$	$13\frac{1}{8}$	$13\frac{1}{8}$	2½	$16\frac{5}{8}$	$16\frac{5}{8}$	20¼
3	—	$13\frac{1}{8}$	$14\frac{1}{8}$	$14\frac{1}{8}$	3	$15\frac{1}{8}$	$18\frac{5}{8}$	23
4	12	$14\frac{5}{8}$	$16\frac{1}{8}$	$17\frac{1}{8}$	4	$18\frac{1}{8}$	$21\frac{5}{8}$	$26\frac{7}{8}$
5	14½	$16\frac{3}{8}$	$18\frac{1}{8}$	$20\frac{1}{8}$	5	$22\frac{1}{8}$	$26\frac{5}{8}$	31¾
6	16½	$18\frac{1}{8}$	$19\frac{5}{8}$	$22\frac{1}{8}$	6	$24\frac{1}{8}$	28	36½
8	20	$22\frac{5}{8}$	$23\frac{5}{8}$	$26\frac{1}{8}$	8	$29\frac{1}{8}$	$33\frac{1}{8}$	$40\frac{7}{8}$
10	25	$25\frac{1}{8}$	$26\frac{5}{8}$	$31\frac{1}{8}$	10	$33\frac{1}{8}$	$39\frac{3}{8}$	$50\frac{7}{8}$
12	28	$28\frac{5}{8}$	$30\frac{1}{8}$	$33\frac{1}{8}$	12	$38\frac{1}{8}$	$45\frac{1}{8}$	$56\frac{7}{8}$
14	31½	—	—	—	14	$40\frac{7}{8}$	50¼	—
16	36½	—	—	—				

Face-to-Face Dimensions of
Flanged Steel Swing Check
Valves

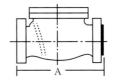

		Raised Face						
	Pressure (lb per sq. in.)					**Pressure (lb per sq. in.)**		
	150	300	400	600		900	1500	2500
Nominal Size (in.)	**Dimension A (in.)**				**Nominal Size (in.)**	**Dimension A (in.)**		
2	8	10½	11½	11½	½	—	—	10⅜
2½	8½	11½	13	13	¾	9	9	10¾
3	9½	12½	14	14	1	10	10	12⅛
3½	10½	13¼	—	—	1¼	11	11	13¾
4	11½	14	16	17	1½	12	12	15⅛
5	13	15¾	—	—	2	14½	14½	17¾
6	14	17½	19½	22	2½	16½	16½	20
8	—	21	23½	26	3	15	18½	22¾
10	—	24½	26½	31	4	18	21½	26½
12	—	28	30	33	5	22	26½	31¼
					6	24	27¾	36
					8	29	32¾	40¼
					10	33	39	50
					12	38	44½	56
					14	40½	49½	—

(continued)

Nominal Size (in.)	Ring-Type Joint Pressure (lb per sq. in.) 150	300	400	600	Nominal Size (in.)	Pressure (lb per sq. in.) 900	1500	2500
	Dimension A (in.)					Dimension A (in.)		
½	$4\frac{11}{16}$	—	$6\frac{7}{16}$	$6\frac{7}{16}$	½	—	—	$10\frac{3}{8}$
¾	$5\frac{1}{8}$	—	$7\frac{1}{2}$	$7\frac{1}{2}$	¾	9	9	$10\frac{3}{4}$
1	$5\frac{1}{2}$	9	$8\frac{1}{2}$	$8\frac{1}{2}$	1	10	10	$12\frac{1}{8}$
1¼	6	$9\frac{1}{2}$	9	9	1¼	11	11	$13\frac{7}{8}$
1½	7	10	$9\frac{1}{2}$	$9\frac{1}{2}$	1½	12	12	$15\frac{1}{4}$
2	$8\frac{1}{2}$	$11\frac{1}{8}$	$11\frac{5}{8}$	$11\frac{5}{8}$	2	$14\frac{5}{8}$	$14\frac{5}{8}$	$17\frac{7}{8}$
2½	9	$12\frac{1}{8}$	$13\frac{1}{8}$	$13\frac{1}{8}$	2½	$16\frac{5}{8}$	$16\frac{5}{8}$	$20\frac{1}{4}$
3	10	$13\frac{1}{8}$	$14\frac{1}{8}$	$14\frac{1}{8}$	3	$15\frac{1}{8}$	$18\frac{5}{8}$	23
4	12	$14\frac{5}{8}$	$16\frac{1}{8}$	$17\frac{1}{8}$	4	$18\frac{1}{8}$	$21\frac{5}{8}$	$26\frac{7}{8}$
5	$13\frac{1}{2}$	$16\frac{3}{8}$	$18\frac{1}{8}$	$20\frac{1}{8}$	5	$22\frac{1}{8}$	$26\frac{5}{8}$	$31\frac{3}{4}$
6	$14\frac{1}{2}$	$18\frac{1}{8}$	$19\frac{5}{8}$	$22\frac{1}{8}$	6	$24\frac{1}{8}$	28	$36\frac{1}{2}$
8	20	$21\frac{5}{8}$	$23\frac{5}{8}$	$26\frac{1}{8}$	8	$29\frac{1}{8}$	$33\frac{1}{8}$	$40\frac{7}{8}$
10	25	$25\frac{1}{8}$	$26\frac{5}{8}$	$31\frac{1}{8}$	10	$33\frac{1}{8}$	$39\frac{3}{8}$	$50\frac{7}{8}$
12	28	$28\frac{5}{8}$	$30\frac{1}{8}$	$33\frac{1}{8}$	12	$38\frac{1}{8}$	$45\frac{1}{8}$	$56\frac{7}{8}$
14	$31\frac{1}{2}$	—	—	—	14	$40\frac{7}{8}$	$50\frac{1}{4}$	—

Reference: Face-to-Face and End-to-End Dimensions of Ferrous Valves American National Standard ANSI B16.10-1973

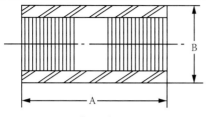

Full coupling

Screwed Couplings
1. All dimensions are in inches.
2. Material, forged carbon steel, conforms to the requirements of SA 105.
3. Threads comply with ANSI standard B2.1-1968.

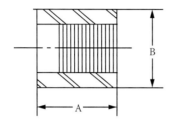

Half coupling

Nominal Pipe Size	Full Coupling				Half Coupling			
	3000 lb		6000 lb		3000 lb		6000 lb	
	Length	Diameter	Length	Diameter	Length	Diameter	Length	Diameter
	A	B	A	B	A	B	A	B
⅛	1¼	¾	1¼	⅞	⅝	¾	⅝	⅞
¼	1⅜	¾	1⅜	1	$^{11}/_{16}$	¾	$^{11}/_{16}$	1
⅜	1½	⅞	1½	1¼	¾	⅞	¾	1¼
½	1⅞	1⅛	1⅞	1½	$^{15}/_{16}$	1⅛	$^{15}/_{16}$	1½
¾	2	1⅜	2	1¾	1	1⅜	1	1¾
1	2⅜	1¾	2⅜	2¼	$1^{3}/_{16}$	1¾	$1^{3}/_{16}$	2¼
1¼	2⅝	2¼	2⅝	2½	$1^{5}/_{16}$	2¼	$1^{5}/_{16}$	2½
1½	3⅛	2½	3⅛	3	$1^{9}/_{16}$	2½	$1^{9}/_{16}$	3
2	3⅜	3	3⅜	3⅝	$1^{11}/_{16}$	3	$1^{11}/_{16}$	3⅝
2½	3⅝	3⅝	3⅝	4¼	$1^{13}/_{16}$	3⅝	$1^{13}/_{16}$	4¼
3	4¼	4¼	4¼	5	2⅛	4¼	2⅛	5
3½	4½	4¾	4½	5¾	2¼	4¾	2¼	5¾
4	4¾	5½	4 3&4	6¼	2⅜	5½	2⅜	6¼

Weight of Pipes and Fittings

Nominal Pipe Size	Designation	Nominal Wall Thickness	Pipe 1 ft.	Elbow 90° L.R.	Elbow 90° S.R.	Elbow 45° L.R.	Return 180° L.R.	Return 180° S.R.	Tee
½	STD	0.109	0.9	0.2		0.1	0.4		0.4
	X STG	0.147	1.1	0.3		0.2	0.5		0.5
	SCH. 160	0.187	1.3						0.4
	XX STG	0.294	1.7						
¾	STD	0.113	1.1	0.2		0.1	0.4		0.5
	X STG	0.154	1.5	0.3		0.2	0.7		0.6
	SCH. 160	0.218	1.9						0.6
	XX STG	0.308	2.4						
1	STD	0.133	1.7	0.4	0.3	0.3	0.8	0.5	0.8
	X STG	0.179	2.2	0.5	0.4	0.3	1.0		0.9
	SCH. 160	0.250	2.8	0.6	0.5	0.3	1.2	0.8	1.0
	XX STG	0.358	3.7	0.8	0.4	0.4	1.5	1.0	1.3
1¼	STD	0.140	2.3	0.6	0.4	0.4	1.3	0.8	1.3
	X STG	0.191	3.0	0.9		0.5	1.8		1.6
	SCH. 160	0.250	3.8	1.0	0.7	0.5	2.0	1.4	2.0
	XX STG	0.382	5.2	1.4	0.9	0.8	2.7	1.8	2.5

(continued)

(continued)

Weight of Pipes and Fittings

Nominal Pipe Size	Designation	Nominal Wall Thickness	Pipe 1 ft.	Elbow 90° L.R.	Elbow 90° S.R.	Elbow 45° L.R.	Return 180° L.R.	Return 180° S.R.	Tee
1½	STD	0.145	2.7	0.9	0.6	0.4	1.9	1.1	2.0
	X STG	0.200	3.6	1.2	0.8	0.7	2.4	1.5	2.3
	SCH. 160	0.281	4.9	1.4	1.2	1.0	3.3	2.4	3.0
	XX STG	0.400	6.4	1.9	1.0	1.1	4.0	2.7	3.4
2	STD	0.154	3.7	1.6	1.0	0.8	3.2	2.0	3.5
	X STG	0.218	5.0	2.2	1.5	1.2	4.4	3.0	4.0
	SCH. 160	0.343	7.5	3.3	2.2	1.6	6.0	4.0	5.0
	XX STG	0.436	9.0	3.5	2.3	2.0	7.5	5.0	6.3
2½	STD	0.203	5.8	3.3	2.1	1.8	6.5	4.3	6.0
	X STG	0.276	7.7	4.0	2.8	2.1	8.0	5.6	7.0
	SCH. 160	0.375	10.0	5.1	3.4	3.0	12.0	6.0	8.0
	XX STG	0.552	13.7	7.0	5.0	3.8	14.0	9.7	10.5
3	STD	0.216	7.6	5.0	3.0	2.6	10.2	6.0	7.0
	X STG	0.300	10.3	6.5	4.3	3.5	13.0	8.5	8.5
	SCH. 160	0.438	14.3	8.5	6.0	4.4	18.0	12.0	10.0
	XX STG	0.600	18.6	11.0	7.3	5.8	22.0	14.6	13.5

3½	STD	0.226	9.1	6.8	4.5	3.5	13.0	9.0	9.0
	X STG	0.318	12.5	8.4	6.0	4.5	16.8	12.0	12.0
	XX STG	0.636	22.9	16.0	11.0	8.5	32.00	22.0	18.0
4	STD	0.237	10.8	9.0	6.3	4.5	18.5	12.5	12.0
	X STG	0.337	15.0	13.5	8.5	6.1	25.0	17.0	15.8
	SCH. 120	0.438	19.0	15.6	10.4	7.8	31.3	20.8	23.5
	SCH. 160	0.531	22.5	18.0	12.0	8.8	40.0	24.0	25.0
	XX STG	0.674	27.5	20.0	13.0	10.8	40.0	27.0	25.0
5	STD	0.258	14.6	15.5	9.6	7.5	30.0	19.0	21.0
	X STG	0.375	20.8	22.0	14.0	10.8	44.0	28.0	26.0
	SCH. 120	0.500	27.0	27.8	18.6	13.9	55.6	37.2	44.5
	SCH. 160	0.625	33.0	32.0	22.0	16.0	65.0	44.0	55.0
	XX STG	0.750	38.6	36.0	24.0	19.0	72.0	48.0	40.0
6	STD	0.280	19.0	24.5	18.0	12.0	50.0	35.0	34.0
	X STG	0.432	28.6	35.0	23.0	17.5	70.0	46.0	40.0
	SCH. 120	0.562	36.4	45.2	30.0	22.6	90.3	60.0	64.0
	SCH. 160	0.718	45.3	57.0	38.0	30.0	120.0	76.0	62.0
	XX STG	0.864	53.2	65.0	44.0	32.0	130.0	87.0	68.0

(continued)

(continued)

Weight of Pipes and Fittings

Nominal Pipe Size	Designation	Nominal Wall Thickness	Pipe 1 ft.	Elbow 90° L.R.	Elbow 90° S.R.	Elbow 45° L.R.	Return 180° L.R.	Return 180° S.R.	Tee
8	SCH. 20	0.250	22.4	36.5	24.4	18.2	73.0	48.8	54.0
	SCH. 30	0.277	24.7	40.9	27.0	20.4	81.9	54.0	57.0
	STD	0.322	28.6	50.0	34.0	23.0	95.0	68.0	55.0
	SCH. 60	0.406	35.6	58.0	39.1	29.4	117.0	78.0	76.0
	X STG	0.500	43.4	71.0	47.5	35.0	142.0	100.0	75.0
	SCH. 100	0.593	50.9	84.0	56.0	42.0	168.0	112.0	97.0
	SCH. 120	0.718	60.6	100.8	66.0	50.4	202.0	133.0	115.0
	SCH. 140	0.812	67.8	111.0	74.0	55.0	222.0	149.0	133.0
	SCH. 160	0.906	74.7	120.0	80.0	62.0	230.0	160.0	152.0
	XX STG	0.875	72.4	118.0	79	60.0	236.0	158.0	148.0
10	SCH. 20	0.250	28.0	56.8	38.2	28.4	114.0	76.4	73.0
	SCH. 30	0.307	34.2	71.4	46.8	35.7	143.0	94.0	81.0
	STD.	0.365	40.5	88.0	58.0	43.0	177.0	115.0	85.0
	X STG	0.500	54.7	107.0	70.0	53.0	215.0	140.0	105.0

Size	Schedule								
(Cont.) 10	SCH. 80	0.592	64.4	133	88	67	267	177	161
	SCH. 100	0.718	77.0	159	106	79	318	212	180
	SCH. 120	0.843	89.2	185	123	92	370	246	215
	SCH. 140	1.000	104.2	214	143	107	428	286	241
	SCH. 160	1.125	116.0	260	174	130	530	348	260
12	SCH. 20	0.250	33.4	82	55	41	164	109	120
	SCH. 30	0.330	43.8	108	72	54	216	145	136
	STD.	0.375	49.6	125	80	62	230	155	120
	SCH. 40	0.406	53.6	132	88	66	264	176	147
	X STG	0.500	65.4	160	104	84	320	218	160
	SCH. 60	0.562	73.2	182	121	91	364	242	226
	SCH. 80	0.687	88.6	219	146	109	439	292	245
	SCH. 100	0.843	108.0	268	177	134	535	354	304
	SCH. 120	1.000	125.5	311	207	155	622	414	353
	SCH. 140	1.125	140.0	347	231	174	694	462	404
	SCH. 160	1.312	161.0	450	300	225	910	600	480

(continued)

(continued)

Weight of Pipes and Fittings

Nominal Pipe Size	Designation	Nominal Wall Thickness	Pipe 1 ft.	Elbow			Return		Tee
				90° L.R.	90° S.R.	45° L.R.	180° L.R.	180° S.R.	
14	SCH. 10	0.250	37.0	106	70	53	212	140	193
	SCH. 20	0.312	46.0	132	87	66	264	175	210
	STD	0.375	55.0	160	105	80	325	210	165
	SCH. 40	0.438	63.0	183	122	91	366	244	252
	X STG	0.500	72.0	205	140	100	400	275	230
	SCH. 60	0.593	85.0	245	163	123	490	326	311
	SCH. 80	0.750	107.0	310	205	154	619	410	369
	SCH. 100	0.937	131.0						
	SCH. 120	1.093	151.0	425		231	850		
	SCH. 140	1.250	171.0						
	SCH. 160	1.406	190.0	572	382	286	1092	764	
16	SCH. 10	0.250	42.0	139	92	69	277	184	201
	SCH. 20	0.312	52.0	172	115	86	344	230	222
	SCH. 30 STD	0.375	63.0	206	132	100	412	260	195
	SCH. 40 X STG	0.500	83.0	276	174	135	550	340	280
	SCH. 60	0.656	108.0	355	236	178	710	472	458

Size	Schedule			450	300	225	900	600	548
(Cont.) 16	SCH. 80	0.843	137						
	SCH. 100	1.031	165						
	SCH. 120	1.218	193						
	SCH. 140	1.438	224						
	SCH. 160	1.593	245						
18	SCH. 10	0.250	47	809	540	405	1618	1080	281
	SCH. 20	0.312	59	176	118	88	352	226	307
	STD	0.375	71	219	146	110	438	292	249
	SCH. 30	0.438	82	260	167	126	510	330	399
	X STG	0.500	93	308	205	154	616	410	332
	SCH. 40	562	105	340	219	167	690	430	525
	SCH. 60	0.750	138	390	259	195	780	518	612
	SCH. 80	0.937	171	494	340	247	989	680	710
	SCH. 100	1.156	208	634	422	317	1268	844	
	SCH. 120	1.375	244						
	SCH. 140	1.562	275						
	SCH. 160	1.781	309						

(continued)

(continued)

Weight of Pipes and Fittings

Nominal Pipe Size	Designation	Nominal Wall Thickness	Pipe 1 ft	Elbow 90° L.R.	Elbow 90° S.R.	Elbow 45° L.R.	Return 180° L.R.	Return 180° S.R.	Tee
20	SCH. 10	0.250	53	217	144	109	434	288	439
	SCH. 20 STD	0.375	79	320	210	160	640	410	342
	SCH. 30 X STD	0.500	105	420	275	206	830	550	480
	SCH. 40	0.593	123	506	338	253	1012	676	706
	SCH. 60	0.812	167	690	457	345	1380	914	834
	SCH. 80	1.031	209	861	573	431	1722	1146	1021
	SCH. 100	1.281	256						
	SCH. 120	1.500	297						
	SCH. 140	1.750	342						
	SCH. 160	1.968	379						
22		0.250	58	262	174	131	524	348	477
		0.312	72						
		0.375	87	394		197	787		414
		0.437	103						
		0.500	115	520		260	1040		550

(Cont.) 22		0.562	129					
		0.625	143					
		0.688	157					
		0.750	170					
24	SCH. 10	0.250	63	314	208	157	627	416	677
	SCH. 20 STD	0.375	95	460	298	238	890	590	528
	X STG	0.500	125	600	392	300	1200	780	610
	SCH. 30	0.562	141	702	470	351	1404	940	977
	SCH. 40	0.687	171	846	564	423	1692	1128	1257
	SCH. 60	0.968	238	1188	783	594	2377	1566	1446
	SCH. 80	1.218	297	1470	977	735	2940	1954	1673
	SCH. 100	1.531	367						
	SCH. 120	1.812	429						
	SCH. 140	2.062	484						
	SCH. 160	2.343	542						

(continued)

(continued)

Nominal Pipe Size	Designation	Nominal Wall Thickness	Pipe 1 ft.	Elbow			Return		Tee
				90° L.R.	90° S.R.	45° L.R.	180° L.R.	180° S.R.	
26		0.250	67						
		0.312	84						
		0.375	103	550		275	1100		770
		0.437	119						
		0.500	136	729		365	1458		875
		0.562	153						
		0.625	169						
		0.688	186						
		0.750	202						
30		0.312	99	612		306	1223		1058
		0.375	119	734	464	367	1465	930	1060
		0.500	158	975	618	488	1950	1235	1200

Weight of Pipes and Fittings

Weight of Flanges

Nominal Pipe Size	150 lb					300 lb				
	Slip On	Weld Neck	Long. Weld Neck	Blind	Studs	Slip On	Weld Neck	Long. Weld Neck	Blind	Studs
½	1.0	2.0		2.0	1.0	1.5	2.0		2.0	1.0
¾	1.5	2.0		2.0	1.0	2.5	3.0		3.0	2.0
1	2.0	2.5	8.0	2.0	1.0	3.0	4.0	10.0	4.0	2.0
1¼	2.5	2.5	10.0	3.0	1.0	4.5	5.0	14.0	6.0	2.0
1½	3.0	4.0	12.0	3.0	1.0	6.5	7.0	17.0	7.0	3.5
2	5.0	6.0	16.0	4.0	1.5	7.0	8.0	19.0	8.0	4.0
2½	8.0	10.0	21.0	7.0	1.5	10.0	12.0	28.0	12.0	7.0
3	9.0	11.5	24.0	9.0	1.5	13.0	16.0	36.0	16.0	7.5
3½	11.0	12.0	31.0	13.0	3.5	16.0	20.0	45.0	21.0	7.5
4	12.0	16.0	47.0	17.0	4.0	21.0	25.0	54.0	27.0	7.5
5	13.0	20.0	57.0	20.0	6.0	26.0	34.0	86.0	35.0	8.0
6	18.0	24.0	77.0	26.0	6.0	35.0	45.0	108.0	50.0	11.5
8	28.0	42.0	103	45.0	6.5	54.0	70.0	150	81.0	18.0
10	37.0	55.0	150	70.0	15.0	77.0	99.0	218	127	38.0
12	60.0	85.0	215	110	15.0	110	142	289	184	49.0
14	77.0	114	221	131	22.0	164	186	342	236	62.0
16	93.0	142	254	170	31.0	220	246	426	307	83.0
18	120	155	278	209	41.0	280	305	493	390	101
20	155	170	324	272	52.0	325	378	575	492	105
22	159	224		333	69.0	433	429		594	157
24	210	260	439	411	71.0	490	545	823	754	174
26	248	270	470	498	93.6	552	615	870	950	239
30	319	375	600	681	112.0	779	858	1130	1403	307

Weight of Flanges

Nominal Pipe Size	400 lb					600 lb				
	Slip On	Weld Neck	Long. Weld Neck	Blind	Studs	Slip On	Weld Neck	Long. Weld Neck	Blind	Studs
½	2.0	3.0		2.0	1.0	2.0	3.0		2.0	1.0
¾	3.0	3.5		3.0	2.0	3.0	3.5		3.0	2.0
1	3.5	4.0	11.0	4.0	2.0	3.5	4.0	11.0	4.0	2.0
1¼	4.5	5.5	14.0	6.0	2.0	4.5	5.5	14.0	6.0	2.0
1½	6.5	8.0	17.0	8.0	3.5	6.5	8.0	17.0	8.0	3.5
2	8.0	10.0	21.0	10.0	4.5	8.0	10.0	21.0	10.0	4.5
2½	12.0	14.0	29.0	15.0	7.5	12.0	14.0	29.0	15.0	8.0
3	15.0	18.0	38.0	20.0	7.7	15.0	18.0	38.0	20.0	8.0
3½	21.0	26.0	48.0	29.0	11.6	21.0	26.0	48.0	29.0	11.6
4	24.0	30.0	67.0	33.0	12.0	33.0	37.0	80.0	41.0	12.5
5	31.0	39.0	90.0	44.0	12.5	63.0	68.0	128	68.0	19.5
6	39.0	49.0	115.0	61.0	19.0	80.0	73.0	158	86.0	30.0
8	63.0	78.0	140	100	30.0	97.0	112.0	215	139	40.0
10	91.0	110.0	230	155	52.0	177	189	324	231	72.0
12	129	160	301	226	69.0	215	226	500	295	91.0
14	191	233	336	310	88.0	259	347	417	378	118
16	253	294	416	398	114	366	481	564	527	152
18	310	360	481	502	139	476	555	654	665	193
20	378	445	563	621	180	612	690	840	855	242
22	464	465		685	205	643	710		962	267
24	539	640	799	936	274	876	977	1100	1175	365
26	616	680	970	1111	307	898	960	1250	1490	398
30	859	940	1230	1596	453	1158	1230	1520	1972	574

Weight of Flanges

Nominal Pipe Size	900 lb					1500 lb				
	Slip On	Weld Neck	Long. Weld Neck	Blind	Studs	Slip On	Weld Neck	Long Weld Neck	Blind	Studs
½	6.0	7.0		4.0	3.2	6.0	7.0		4.0	3.2
¾	6.0	7.0		6.0	3.3	6.0	7.0		6.0	3.3
1	7.5	8.5	15.0	9.0		7.5	8.5	15.0	9.0	6.0
1¼	10.0	10.0	18.0	10.0		10.0	10.0	18.0	10.0	6.0
1½	14.0	14.0	23.0	14.0		14.0	14.0	23.0	14.0	9.0
2	25.0	24.0	44.0	25.0		25.0	24.0	44.0	25.0	12.5
2½	36.0	36.0	65.0	35.0	19.0	36.0	36.0	72.0	35.0	19.0
3	31.0	29.0	72.0	32.0	12.5	48.0	48.0	84.0	48.0	25.0
3½										
4	53.0	51.0	98.0	54.0	25.0	73.0	69.0	118	73.0	34.0
5	83.0	86.0	143	87.0	33.0	132.0	132.0	195	142	60.0
6	108.0	110.0	199	113	40.0	164	164	235	159	76.0
8	172	187	310	197	69.0	258	273	366	302	121
10	245	268	385	290	95.0	436	454	610	507	184
12	326	372	667	413	124	667	690	1028	775	306
14	380	562	558	494	159	Weights on application	940	1030	975	425
16	459	685	670	619	199		1250	1335	1300	570
18	647	924	949	880	299		1625	1750	1750	770
20	792	1164	1040	1107	361		2050	2130	2225	1010
22										
24	1480	2107	1775	2099	687	1525	3325	3180	3625	1560
26	1450	1650	1650	2200	765	1575	1575		2200	
30	1990	2290	2200	3025	1074	2075	2150		3025	

Weight of Flanges

Nominal Pipe Size	2500 lb									
	Slip On	Weld Neck	Long. Weld Neck	Blind	Studs	Slip On	Weld Neck	Long. Weld Neck	Blind	Studs
½	7.0	8.0		7.0	3.4					
¾	9.0	9.0		10.0	3.6					
1	12.0	13.0	20.0	12.0	6.0					
1¼	18.0	20.0	30.0	18.0	9.0					
1½	25.0	28.0	38.0	25.0	12.0					
2	38.0	42.0	55.0	39.0	21.0					
2½	55.0	52.0	85.0	56.0	27.0					
3	83.0	94.0	125.0	86.0	37.0					
3½										
4	127	146	185	133	61					
5	210	244	300	223	98					
6	323	378	450	345	145					
8	485	576	600	533	232					
10	925	1068	1150	1025	445					
12	1300	1608	1560	1464	622					
14										
16										
18										
20										
22										
24										
26										
30										

Weight of Plates
(Pounds per Linear Foot)

Width (in.)	Thickness (in.)													
	³⁄₁₆	¼	⁵⁄₁₆	³⁄₈	⁷⁄₁₆	½	⁹⁄₁₆	⅝	¹¹⁄₁₆	¾	¹³⁄₁₆	⅞	¹⁵⁄₁₆	1
¼	0.16	0.21	0.27	0.32	0.37	0.43	0.48	0.53	0.58	0.64	0.69	0.74	0.80	0.85
½	0.32	0.43	0.53	0.64	0.74	0.85	0.96	1.06	1.17	1.28	1.38	1.49	1.59	1.70
¾	0.48	0.64	0.80	0.96	1.12	1.28	1.43	1.59	1.75	1.91	2.07	2.23	2.39	2.55
1	0.64	0.85	1.06	1.28	1.49	1.70	1.91	2.13	2.34	2.55	2.76	2.98	3.19	3.40
1¼	0.80	1.06	1.33	1.59	1.86	2.13	2.39	2.66	2.92	3.19	3.45	3.72	3.98	4.25
1½	0.96	1.28	1.59	1.91	2.23	2.55	2.87	3.19	3.51	3.83	4.14	4.46	4.78	5.10
1¾	1.12	1.49	1.86	2.23	2.60	2.98	3.35	3.72	4.09	4.46	4.83	5.21	5.58	5.95
2	1.28	1.70	2.13	2.55	2.98	3.40	3.83	4.25	4.68	5.10	5.53	5.95	6.38	6.80
2¼	1.43	1.91	2.39	2.87	3.35	3.83	4.30	4.78	5.26	5.74	6.22	6.69	7.17	7.65
2½	1.59	2.13	2.66	3.19	3.72	4.25	4.78	5.31	5.84	6.38	6.91	7.44	7.97	8.50
2¾	1.75	2.34	2.92	3.51	4.09	4.68	5.26	5.84	6.43	7.01	7.60	8.18	8.77	9.35
3	1.91	2.55	3.19	3.83	4.46	5.10	5.74	6.38	7.01	7.65	8.29	8.93	9.56	10.2
3¼	2.07	2.76	3.45	4.14	4.83	5.53	6.22	6.91	7.60	8.29	8.98	9.67	10.4	11.1
3½	2.23	2.98	3.72	4.46	5.21	5.95	6.69	7.44	8.18	8.93	9.67	10.4	11.2	11.9
3¾	2.39	3.19	3.98	4.78	5.58	6.38	7.17	7.97	8.77	9.56	10.4	11.2	12.0	12.8
4	2.55	3.40	4.25	5.10	5.95	6.80	7.65	8.50	9.35	10.2	11.1	11.9	12.8	13.6
4¼	2.71	3.61	4.52	5.42	6.32	7.23	8.13	9.03	9.93	10.8	11.7	12.6	13.6	14.5
4½	2.87	3.83	4.78	5.74	6.69	7.65	8.61	9.56	10.5	11.5	12.4	13.4	14.3	15.3
4¾	3.03	4.04	5.05	6.06	7.07	8.08	9.08	10.1	11.1	12.1	13.1	14.1	15.1	16.2
5	3.19	4.25	5.31	6.38	7.44	8.50	9.56	10.6	11.7	12.8	13.8	14.9	15.9	17.0
5¼	3.35	4.46	5.58	6.69	7.81	8.93	10.0	11.2	12.3	13.4	14.5	15.6	16.7	17.9
5½	3.51	4.68	5.84	7.01	8.18	9.35	10.5	11.7	12.9	14.0	15.2	16.4	17.5	18.7
5¾	3.67	4.89	6.11	7.33	8.55	9.78	11.0	12.2	13.4	14.7	15.9	17.1	18.3	19.6

(continued)

(continued)

Weight of Plates
(Pounds per Linear Foot)

Width (in.)	| Thicknes (in.)													
	3/16	1/4	5/16	3/8	7/16	1/2	9/16	5/8	11/16	3/4	13/16	7/8	15/16	1
6	3.83	5.10	6.38	7.65	8.93	10.2	11.5	12.8	14.0	15.3	16.6	17.9	19.1	20.4
6¼	3.98	5.31	6.64	7.97	9.30	10.6	12.0	13.3	14.6	15.9	17.3	18.6	19.9	21.3
6½	4.14	5.53	6.91	8.29	9.67	11.1	12.4	13.8	15.2	16.6	18.0	19.3	20.7	22.1
6¾	4.30	5.74	7.17	8.61	10.0	11.5	12.9	14.3	15.8	17.2	18.7	20.1	21.5	23.0
7	4.46	5.95	7.44	8.93	10.4	11.9	13.4	14.9	16.4	17.9	19.3	20.8	22.3	23.8
7¼	4.62	6.16	7.70	9.24	10.8	12.3	13.9	15.4	17.0	18.5	20.0	21.6	23.1	24.7
7½	4.78	6.38	7.97	9.56	11.2	12.8	14.3	15.9	17.5	19.1	20.7	22.3	23.9	25.5
7¾	4.94	6.59	8.23	9.98	11.5	13.2	14.8	16.5	18.1	19.8	21.4	23.1	24.7	26.4
8	5.10	6.80	8.50	10.2	11.9	13.6	15.3	17.0	18.7	20.4	22.1	23.8	25.5	27.2
8¼	5.26	7.01	8.77	10.5	12.3	14.0	15.8	17.5	19.3	21.0	22.8	24.5	26.3	28.1
8½	5.42	7.23	9.03	10.8	12.6	14.5	16.3	18.1	19.9	21.7	23.5	25.3	27.1	28.9
8¾	5.58	7.44	9.30	11.2	13.0	14.9	16.7	18.6	20.5	22.3	24.2	26.0	27.9	29.8
9	5.74	7.65	9.56	11.5	13.4	15.3	17.2	19.1	21.0	23.0	24.9	26.8	28.7	30.6
9¼	5.90	7.86	9.83	11.8	13.8	15.7	17.7	19.7	21.6	23.6	5.6	27.5	29.5	31.5
9½	6.06	8.08	10.1	12.1	14.1	16.2	18.2	20.2	22.2	24.2	26.2	28.3	30.3	32.3
9¾	6.22	8.29	10.4	12.4	14.5	16.6	18.7	20.7	22.8	24.9	26.9	29.0	31.1	33.2
10	6.38	8.50	10.6	12.8	14.9	17.0	19.1	21.3	23.4	25.5	27.6	29.8	31.9	34.0

10¼	6.53	8.71	10.9	13.1	15.3	17.4	19.6	21.8	24.0	26.1	28.3	30.5	32.7	34.9
10½	6.69	8.93	11.2	13.4	15.6	17.9	20.1	22.3	24.5	26.8	29.0	31.2	33.5	35.7
10¾	6.85	9.14	11.4	13.7	16.0	18.3	20.6	22.8	25.1	27.4	29.7	32.0	34.3	36.6
11	7.01	9.35	11.7	14.0	16.4	18.7	21.0	23.4	25.7	28.1	30.4	32.7	35.1	37.4
11¼	7.17	9.56	12.0	14.3	16.7	19.1	21.5	23.9	26.3	28.7	31.1	33.5	35.9	38.3
11½	7.33	9.78	12.2	14.7	17.1	19.6	22.0	24.4	26.9	29.3	31.8	34.2	36.7	39.1
11¾	7.49	9.99	12.5	15.0	17.5	20.0	22.5	25.0	27.5	30.0	32.5	35.0	37.5	40.0
12	7.65	10.2	12.8	15.3	17.9	20.4	23.0	25.5	28.1	30.6	33.2	35.7	38.3	40.8
12½	7.97	10.6	13.3	15.9	18.6	21.3	23.9	26.6	29.2	31.9	34.5	37.2	39.8	42.5
13	8.29	11.1	13.8	16.6	19.3	22.1	24.9	27.6	30.4	33.2	35.9	38.7	41.4	44.2
13½	8.61	11.5	14.3	17.2	20.1	23.0	25.8	28.7	31.6	34.4	37.3	40.2	43.0	45.9
14	8.93	11.9	14.9	17.9	20.8	23.8	26.8	29.8	32.7	35.7	38.7	41.7	44.6	47.6
14½	9.24	12.3	15.4	18.5	21.6	24.7	27.7	30.8	33.9	37.0	40.1	43.1	46.2	49.3
15	9.56	12.8	15.9	19.1	22.3	25.5	28.7	31.9	35.1	38.3	41.4	44.6	47.8	51.0
15½	9.88	13.2	16.5	19.8	23.1	26.4	29.6	32.9	36.2	39.5	42.8	46.1	49.4	52.7
16	10.2	13.6	17.0	20.4	23.8	27.2	30.6	34.0	37.4	40.8	44.2	47.6	51.0	54.4
16½	10.5	14.0	17.5	21.0	24.5	28.1	31.6	35.1	38.6	42.1	45.6	49.1	52.6	56.1
17	10.8	14.5	18.1	21.7	25.3	28.9	32.5	36.1	39.7	43.4	47.0	50.6	54.2	57.8
17½	11.2	14.9	18.6	22.3	26.0	29.8	33.5	37.2	40.9	44.6	48.3	52.1	55.8	59.5
18	11.5	15.3	19.1	23.0	26.8	30.6	34.4	38.3	42.1	45.9	49.9	53.6	57.4	61.2
18½	11.8	15.7	19.7	23.6	27.5	31.5	35.4	39.3	43.2	47.2	51.1	55.0	59.0	62.9
19	12.1	16.2	20.2	24.2	28.3	32.3	36.3	40.4	44.4	48.5	52.5	56.5	60.6	64.6
19½	12.4	16.6	20.7	24.9	29.0	33.2	37.3	41.4	45.6	49.7	53.9	58.0	62.2	66.3

(continued)

(continued)

Weight of Plates
(Pounds per Linear Foot)

Thickness (in.)

Width (in.)	$\frac{3}{16}$	$\frac{1}{4}$	$\frac{5}{16}$	$\frac{3}{8}$	$\frac{7}{16}$	$\frac{1}{2}$	$\frac{9}{16}$	$\frac{5}{8}$	$\frac{11}{16}$	$\frac{3}{4}$	$\frac{13}{16}$	$\frac{7}{8}$	$\frac{15}{16}$	1
20	12.8	17.0	21.3	25.5	29.8	34.0	38.3	42.5	46.8	51.0	55.3	59.5	63.8	68.0
20½	13.1	17.4	21.8	26.1	30.5	34.9	39.2	43.6	47.9	52.3	56.6	61.0	65.3	69.7
21	13.4	17.9	22.3	26.8	31.2	35.7	40.2	44.6	49.1	53.6	58.0	62.5	66.9	71.4
21½	13.7	18.3	22.8	27.4	32.0	36.6	41.1	45.7	50.3	54.8	59.4	64.0	68.5	73.1
22	14.0	18.7	23.4	28.1	32.7	37.4	42.1	46.8	51.4	56.1	60.8	65.5	70.1	74.8
22½	14.3	19.1	23.9	28.7	33.5	38.3	43.0	47.8	52.6	57.4	62.2	66.9	71.7	76.5
23	14.7	19.6	24.4	29.3	34.2	39.1	44.0	48.9	53.8	58.7	63.5	68.4	73.3	78.2
23½	15.0	20.0	25.0	30.0	35.0	40.0	44.9	49.9	54.9	59.9	64.9	69.9	74.9	79.9
24	15.3	20.4	25.5	30.6	35.7	40.8	45.9	51.0	56.1	61.2	66.3	71.4	76.5	81.6
25	15.9	21.3	26.6	31.9	37.2	42.5	47.8	53.1	58.4	63.8	69.1	74.4	79.7	85.0
26	16.6	22.1	27.6	33.2	38.7	44.2	49.7	55.3	60.8	66.3	71.8	77.4	82.9	88.4
27	17.2	23.0	28.7	34.4	40.2	45.9	51.6	57.5	63.1	68.9	74.6	80.3	86.1	91.8
28	17.9	23.8	29.8	35.7	41.7	47.6	53.6	59.5	65.5	71.4	77.4	83.3	89.3	95.2
29	18.5	24.7	30.8	37.0	43.1	49.3	55.5	61.6	67.8	74.0	80.1	86.3	92.4	98.6
30	19.1	25.5	31.9	38.3	44.6	51.0	57.4	63.8	70.1	76.5	82.9	89.3	95.6	102
31	19.8	26.4	32.9	39.5	46.1	52.7	59.3	65.9	72.5	79.1	85.6	92.2	98.8	105
32	20.4	27.2	34.0	40.8	47.6	54.4	61.2	68.0	74.8	81.6	88.4	95.2	102	109

33	21.0	28.1	35.1	42.1	49.1	56.1	63.1	70.1	77.1	84.2	91.2	98.2	105	112
34	21.7	28.9	36.1	43.4	50.6	57.8	65.0	72.3	79.5	86.7	93.9	101	108	116
35	22.3	29.8	37.2	44.6	52.1	59.5	66.9	74.4	81.8	89.3	96.1	104	112	119
36	23.0	30.6	38.3	45.9	53.6	61.2	68.9	76.5	84.2	91.8	99.5	107	115	122
37	23.6	31.5	39.3	47.2	55.0	62.9	70.8	78.6	86.5	94.4	102	110	118	126
38	24.2	32.3	40.4	48.5	56.5	64.6	72.7	80.8	88.8	96.9	105	113	121	129
39	24.9	33.2	41.4	49.7	58.0	66.3	74.6	82.9	91.2	99.5	108	116	124	133
40	25.5	34.0	42.5	51.0	59.5	68.0	76.5	85.0	93.5	102	111	119	128	136
41	26.1	34.9	43.6	52.3	61.0	69.7	78.4	87.1	95.8	105	113	122	131	139
42	26.8	35.7	44.6	53.6	62.5	71.4	80.3	89.3	98.2	107	116	125	134	143
43	27.4	36.6	45.7	54.8	64.0	73.1	82.2	91.4	101	110	119	128	137	146
44	28.1	37.4	46.8	56.1	65.5	74.8	84.2	93.5	103	112	122	131	140	150
45	28.7	38.3	47.8	57.4	66.9	76.5	86.1	95.6	105	115	124	134	143	153
46	29.3	39.1	48.9	58.7	68.4	78.2	88.0	97.8	108	117	127	137	147	156
47	30.0	40.0	49.9	59.9	69.9	79.9	89.9	99.9	110	120	130	140	150	160
48	30.6	40.8	51.0	61.2	71.4	81.6	91.8	102	112	122	133	143	153	163
49	31.2	41.7	52.1	62.5	72.9	83.3	93.7	104	115	125	135	146	156	167
50	21.9	42.5	53.1	63.8	74.4	85.0	95.6	106	117	128	138	149	159	170
51	32.5	43.4	54.2	65.0	75.9	86.7	97.5	108	119	130	141	152	163	173
52	33.2	44.2	55.3	66.3	77.4	88.4	99.5	111	122	133	144	155	166	177
53	33.8	45.1	56.3	67.6	78.8	90.1	101	113	124	135	146	158	169	180
54	34.4	45.9	57.4	68.9	80.3	91.8	103	115	126	138	149	161	172	184

(continued)

(continued)

Weight of Plates
(Pounds per Linear Foot)

Width (in.)	Thickness (in.)													
	$3/16$	$1/4$	$5/16$	$3/8$	$7/16$	$1/2$	$9/16$	$5/8$	$11/16$	$3/4$	$13/16$	$7/8$	$15/16$	1
55	35.1	46.8	58.4	70.1	81.8	93.5	105	117	129	140	152	164	175	187
56	35.7	47.6	59.5	71.4	83.3	95.2	107	119	131	143	155	167	179	190
57	36.3	48.5	60.6	72.7	84.8	96.9	109	121	133	145	158	170	182	194
58	37.0	49.5	61.6	74.0	86.3	98.6	111	123	136	148	160	173	185	197
59	37.6	50.2	62.7	75.2	87.8	100	113	125	138	151	163	176	188	201
60	38.3	51.0	63.8	76.5	89.3	102	115	128	140	153	166	179	191	204
61	38.9	51.9	64.8	77.8	90.7	104	117	130	143	156	169	182	194	207
62	39.5	52.7	65.9	79.1	92.2	105	119	132	145	158	171	185	198	211
63	40.2	53.6	66.9	80.3	93.7	107	121	134	147	161	174	187	201	214
64	20.8	54.4	68.0	81.6	95.2	109	122	136	150	163	177	190	204	218
65	41.4	55.3	69.1	82.9	96.7	111	124	138	152	166	180	193	207	221
66	42.1	56.1	70.1	84.2	98.2	112	126	140	154	168	182	196	210	224
67	42.7	57.0	71.2	85.4	99.7	114	128	142	157	171	185	199	214	228
68	43.4	57.8	72.3	86.7	101	116	130	145	159	173	188	202	217	231
69	44.0	58.7	73.3	88.0	103	117	132	147	161	176	191	205	220	235
70	44.6	59.5	74.4	89.3	104	119	134	149	164	179	193	208	223	238
71	45.3	60.4	75.4	90.5	106	121	136	151	166	181	196	211	226	241
72	45.9	61.2	76.5	91.8	107	122	138	153	168	184	199	214	230	245

73	46.5	62.1	77.6	93.1	109	124	140	155	171	186	202	217	233	248
74	47.2	62.9	78.6	94.4	110	126	142	157	173	189	204	220	236	252
75	47.8	63.8	79.7	95.6	112	128	143	159	175	191	207	223	239	255
76	48.5	64.6	80.8	96.9	113	129	145	162	178	194	210	226	242	258
77	49.1	65.5	81.8	98.2	115	131	147	164	180	196	213	229	245	262
78	49.7	66.3	82.9	99.5	116	133	149	166	182	199	216	232	249	265
79	50.4	67.2	83.9	101	118	134	151	168	185	202	218	235	252	269
80	51.0	68.0	85.0	102	119	136	153	170	187	204	221	238	255	272
81	51.6	68.9	86.1	103	121	138	155	172	189	207	224	241	258	275
82	52.3	69.7	87.1	105	122	139	157	174	192	209	227	244	261	279
83	52.9	70.6	88.2	106	124	141	159	176	194	212	229	247	265	282
84	53.6	71.4	89.3	107	125	143	161	179	196	214	232	250	268	286
85	54.2	72.3	90.3	108	126	145	163	181	199	217	235	253	271	289
86	54.8	73.1	91.4	110	128	146	165	183	201	219	238	256	274	292
87	55.5	74.0	92.4	111	129	148	166	185	203	222	240	259	277	296
88	56.1	74.8	93.5	112	131	150	168	187	206	224	243	262	281	299
89	56.7	75.7	94.6	114	132	151	170	189	208	227	246	265	284	303
90	57.4	76.5	95.6	115	134	153	172	191	210	230	249	268	287	306
91		77.4	96.7	116	135	155	174	193	213	232	251	271	290	309
92		78.2	97.8	117	137	156	176	196	215	235	254	274	293	313
93		79.1	98.8	119	138	158	178	198	217	237	257	277	296	316
94		79.9	99.9	120	140	160	180	200	220	240	260	280	300	320

(continued)

(continued)

Weights of Plates
(Pounds per Linear Foot)

Thickness (in.)

Width (in.)	3/16	1/4	5/16	3/8	7/16	1/2	9/16	5/8	11/16	3/4	13/16	7/8	15/16	1
95		80.8	101	121	141	162	182	202	222	242	262	283	303	323
96		81.6	102	122	143	163	184	204	224	245	265	286	306	326
98		83.3	104	125	146	167	187	208	229	250	271	292	312	333
100		85.0	106	128	149	170	191	213	234	255	276	298	319	340
102		86.7	108	130	152	173	195	217	238	260	282	304	325	347
104		88.4	111	133	155	177	199	221	243	265	287	309	332	354
106		90.1	113	135	158	180	203	225	248	270	293	315	338	360
108		91.8	115	138	161	184	207	230	253	275	298	321	344	367
110		93.5	117	140	164	187	210	234	257	281	304	327	351	374
112		95.2	119	143	167	190	214	238	262	286	309	333	357	381
114		96.9	121	145	170	194	218	242	267	291	315	339	363	388
116		98.6	123	148	173	197	222	247	271	296	321	345	370	394
118		100	125	151	176	201	226	251	276	301	326	351	376	401
120		102	128	153	179	204	230	255	281	306	332	357	383	408
122		104	130	156	182	207	233	259	285	311	337	363	389	415
124		105	132	158	185	211	237	264	290	316	343	369	395	422
126		107	134	161	187	214	241	268	295	321	348	375	402	428
128		109	136	163	190	218	245	272	299	326	354	381	408	435

Conversion Table—Length
Inches to Millimeters (1 in. = 25.4 mm)

Inches	0	$\frac{1}{16}$	$\frac{1}{8}$	$\frac{3}{16}$	$\frac{1}{4}$	$\frac{5}{16}$	$\frac{3}{8}$	$\frac{7}{16}$	$\frac{1}{2}$	$\frac{9}{16}$	$\frac{5}{8}$	$\frac{11}{16}$	$\frac{3}{4}$	$\frac{13}{16}$	$\frac{7}{8}$	$\frac{15}{16}$
0	0.0	1.6	3.2	4.8	6.4	7.9	9.5	11.1	12.7	14.3	15.9	17.5	19.1	20.6	22.2	23.8
1	25.4	27.0	28.6	30.2	31.8	33.3	34.9	36.5	38.1	39.7	41.3	42.9	44.5	46.0	47.6	49.2
2	50.8	52.4	54.0	55.6	57.2	58.7	60.3	61.9	63.5	65.1	66.7	68.3	69.9	71.4	73.0	74.6
3	76.2	77.8	79.4	81.0	82.6	84.1	85.7	87.3	88.9	90.5	92.1	93.7	95.3	96.8	98.4	100.0
4	101.6	103.2	104.8	106.4	108.0	109.5	111.1	112.7	114.3	115.9	117.5	119.1	120.7	122.2	123.8	125.4
5	127.0	128.6	130.2	131.8	133.4	134.9	136.5	138.1	139.7	141.3	142.9	144.5	146.1	147.6	149.2	150.8
6	152.4	154.0	155.6	157.2	158.8	160.3	161.9	163.5	165.1	166.7	168.3	169.9	171.5	173.0	174.6	176.2
7	177.8	179.4	181.0	182.6	184.2	185.7	187.3	188.9	190.5	192.1	193.7	195.3	196.9	198.4	200.0	201.6
8	203.2	204.8	206.4	208.0	209.6	211.1	212.7	214.3	215.9	217.5	219.1	220.7	222.3	223.8	225.4	227.0
9	228.6	230.2	231.8	233.4	235.0	236.5	238.1	239.7	241.3	242.9	244.5	246.1	247.7	249.2	250.8	252.4
10	254.0	255.6	257.2	258.8	260.4	261.9	263.5	265.1	266.7	268.3	269.9	271.5	273.1	274.6	276.2	277.8
11	279.4	281.0	282.6	284.2	285.8	287.3	288.9	290.5	292.1	293.7	295.3	296.9	298.5	300.0	301.6	303.2
12	304.8	306.4	308.0	309.6	311.2	312.7	314.3	315.9	317.5	319.1	320.7	322.3	323.9	325.4	327.4	354.0
13	330.2	331.8	333.4	335.0	336.6	338.1	339.7	341.3	342.9	344.5	346.1	347.7	349.3	350.8	352.4	354.0
14	355.6	357.2	358.8	360.4	362.0	363.5	365.1	366.7	368.3	369.9	371.5	373.1	374.7	376.2	377.8	379.4
15	381.0	382.6	384.2	385.8	387.4	388.9	390.5	392.1	393.7	395.3	396.9	398.5	400.1	401.6	403.2	404.8
16	406.4	408.0	409.6	411.2	412.8	414.3	415.9	417.5	419.1	420.7	422.3	423.9	425.5	427.0	428.6	430.2
17	431.8	433.4	435.0	436.6	438.2	439.7	441.3	442.9	444.5	446.1	447.7	449.3	450.9	452.4	454.0	455.6
18	457.2	458.8	460.4	462.0	463.6	465.1	466.7	468.3	469.9	471.5	473.1	474.7	476.3	477.8	479.4	481.0
19	482.6	484.2	485.8	487.4	489.0	490.5	492.1	493.7	495.3	496.9	498.5	500.1	501.7	503.2	504.8	506.4
20	508.0	509.6	511.2	512.8	514.4	515.9	517.5	519.1	520.7	522.3	523.9	525.5	527.1	528.6	530.2	531.8
21	533.4	535.0	536.6	538.2	539.8	541.3	542.9	544.5	546.1	547.7	549.3	550.9	552.5	554.0	555.6	557.2
22	558.8	560.4	562.0	563.6	565.2	566.7	568.3	569.9	571.5	573.1	574.7	576.3	577.9	579.4	581.0	582.6
23	584.2	585.8	587.4	589.0	590.6	592.1	593.7	595.3	596.9	598.5	600.1	601.7	603.3	604.8	606.4	608.0
24	609.6	611.2	612.8	614.4	616.0	617.5	619.1	620.7	622.3	623.9	625.5	627.1	628.7	630.2	631.8	633.4

(continued)

Appendix F: Dimension of Flanges and Fittings

(continued)

Conversion Table—Length
Inches to Millimeters (1 in. = 25.4 mm)

Inches	0	1/16	1/8	3/16	1/4	5/16	3/8	7/16	1/2	9/16	5/8	11/16	3/4	13/16	7/8	15/16
25	635.0	636.6	638.2	639.8	641.4	642.9	644.5	646.1	647.7	649.3	650.9	652.5	654.1	655.6	657.2	658.8
26	660.4	662.0	663.6	665.2	666.8	668.3	669.9	671.5	673.1	674.7	676.3	677.9	679.5	681.0	682.6	684.2
27	685.8	687.4	689.0	690.6	692.2	693.7	695.3	696.9	698.5	700.1	701.7	703.3	704.9	706.4	708.0	709.6
28	711.2	712.8	714.4	716.0	717.6	719.1	720.7	722.3	723.9	725.5	727.1	728.7	730.3	731.8	733.4	735.0
29	736.6	738.2	739.8	741.4	743.0	744.5	746.1	747.7	749.3	750.9	752.5	754.1	755.7	757.2	758.8	760.4
30	762.0	763.6	765.2	766.8	768.4	769.9	771.5	773.1	774.7	776.3	777.9	779.5	781.1	782.6	784.2	785.8
31	787.4	789.0	790.6	792.2	793.8	795.3	796.9	798.5	800.1	801.7	803.3	804.9	806.5	808.0	809.6	811.2
32	812.8	814.4	816.0	817.6	819.2	820.7	822.3	823.9	825.5	827.1	828.7	830.3	831.9	833.4	835.0	836.6
33	838.2	839.8	841.4	843.0	844.6	846.1	847.7	849.3	850.9	852.5	854.1	855.7	857.3	858.8	860.4	862.0
34	863.6	865.2	866.8	868.4	870.0	871.5	873.1	874.7	876.3	877.9	879.5	881.1	882.7	884.2	885.8	887.4
35	889.0	890.6	892.2	893.8	895.4	896.9	898.5	900.1	901.7	903.3	904.9	906.5	908.1	909.6	911.2	912.8
36	914.4	916.0	917.6	919.2	920.8	922.3	923.9	925.5	927.1	928.7	930.3	931.9	933.5	935.0	936.6	938.2
37	939.8	941.4	943.0	944.6	946.2	947.7	949.3	950.9	952.5	954.1	955.7	957.3	958.9	960.4	962.0	963.6
38	965.2	966.8	968.4	970.0	971.6	973.1	974.7	976.3	977.9	979.5	981.1	982.7	984.3	985.8	987.4	989.0
39	990.6	992.2	993.8	995.4	997.0	998.5	1000.1	1001.7	1003.3	1004.9	1006.5	1008.1	1009.7	1011.2	1012.8	1014.4
40	1016.0	1017.6	1019.2	1020.8	1022.4	1023.9	1025.5	1027.1	1028.7	1030.3	1031.9	1033.5	1035.1	1036.6	1038.2	1039.8
41	1041.4	1043.0	1044.6	1046.2	1047.8	1049.3	1050.9	1052.5	1054.1	1055.7	1057.3	1058.9	1060.5	1062.0	1063.6	1065.2
42	1066.8	1068.4	1070.0	1071.6	1073.2	1074.7	1076.3	1077.9	1079.5	1081.1	1082.7	1084.3	1085.9	1087.4	1089.0	1090.6
43	1092.2	1093.8	1095.4	1097.0	1098.6	1100.1	1101.7	1103.3	1104.9	1106.5	1108.1	1109.7	1111.3	1112.8	1114.4	1116.0
44	1117.6	1119.2	1120.8	1122.4	1124.0	1125.5	1127.1	1128.7	1130.3	1131.9	1133.5	1135.1	1136.7	1138.2	1139.8	1141.4
45	1143.0	1144.6	1146.2	1147.8	1149.4	1150.9	1152.5	1154.1	1155.7	1157.3	1158.9	1160.5	1162.1	1163.6	1165.2	1166.8
46	1168.4	1170.0	1171.6	1173.2	1174.8	1176.3	1177.9	1179.5	1181.1	1182.7	1184.3	1185.9	1187.5	1189.0	1190.6	1192.2
47	1193.8	1195.4	1197.0	1198.6	1200.2	1201.7	1203.3	1204.9	1206.5	1208.1	1209.7	1211.3	1212.9	1214.4	1216.0	1217.6
48	1219.2	1220.8	1222.4	1224.0	1225.6	1227.1	1228.7	1230.3	1231.9	1233.5	1235.1	1236.7	1238.3	1239.8	1241.4	1243.0
49	1244.6	1246.2	1247.8	1249.4	1251.0	1252.5	1254.1	1255.7	1257.3	1258.9	1260.5	1262.1	1263.7	1265.2	1266.8	1268.4
50	1270.0	1271.6	1273.2	1274.8	1276.4	1277.9	1279.5	1281.1	1282.7	1284.3	1285.9	1287.5	1289.1	1290.6	1292.2	1293.8

Conversion Table—Length
Millimeters to Inches (1 mm = 0.0394 in.)

Millimeters	0	1	2	3	4	5	6	7	8	9	Millimeters
0	0.00	0.039	0.079	0.118	0.157	0.197	0.236	0.276	0.315	0.354	0
10	0.39	0.43	0.47	0.51	0.55	0.59	0.63	0.67	0.71	0.75	10
20	0.79	0.83	0.87	0.91	0.94	0.98	1.02	1.06	1.10	1.14	20
30	1.18	1.22	1.26	1.30	1.34	1.38	1.42	1.46	1.50	1.54	30
40	1.57	1.61	1.65	1.69	1.73	1.77	1.81	1.85	1.89	1.93	40
50	1.97	2.01	2.05	2.09	2.13	2.17	2.20	2.24	2.28	2.32	50
60	2.36	2.40	2.44	2.48	2.52	2.56	2.60	2.64	2.68	2.72	60
70	2.76	2.80	2.83	2.87	2.91	2.95	2.99	3.03	3.07	3.11	70
80	3.15	3.19	3.23	3.27	3.31	3.35	3.39	3.43	3.46	3.50	80
90	3.54	3.58	3.62	3.66	3.70	3.74	3.78	3.82	3.86	3.90	90
100	3.94	3.98	4.02	4.06	4.09	4.13	4.17	4.21	4.25	4.29	100
110	4.33	4.37	4.41	4.45	4.49	4.53	4.57	4.61	4.65	4.69	110
120	4.72	4.76	4.80	4.84	4.88	4.92	4.96	5.00	5.04	5.08	120
130	5.12	5.16	5.20	5.24	5.28	5.31	5.35	5.39	5.43	5.47	130
140	5.51	5.55	5.59	5.63	5.67	5.71	5.75	5.79	5.83	5.87	140
150	5.91	5.94	5.98	6.02	6.06	6.10	6.14	6.18	6.22	6.26	150

(continued)

(continued)

Conversion Table—Length
Millimeters to Inches (1 mm = 0.0394 in.)

Millimeters	0	1	2	3	4	5	6	7	8	9	Millimeters
160	6.30	6.34	6.38	6.42	6.46	6.50	6.54	6.57	6.61	6.65	160
170	6.69	6.73	6.77	6.81	6.85	6.89	6.93	6.97	7.01	7.05	170
180	7.09	7.13	7.17	7.20	7.24	7.28	7.32	7.36	7.40	7.44	180
190	7.48	7.52	7.56	7.60	7.64	7.68	7.72	7.76	7.80	7.83	190
200	7.87	7.91	7.95	7.99	8.03	8.07	8.11	8.15	8.19	8.23	200
210	8.27	8.31	8.35	8.39	8.43	8.46	8.50	8.54	8.58	8.62	210
220	8.66	8.70	8.74	8.78	8.82	8.86	8.90	8.94	8.98	9.02	220
230	9.06	9.09	9.13	9.17	9.21	9.25	9.29	9.33	9.37	9.41	230
240	9.45	9.49	9.53	9.57	9.61	9.65	9.69	9.72	9.76	9.80	240
250	9.84	9.88	9.92	9.96	10.00	10.04	10.08	10.12	10.16	10.20	250
260	10.24	10.28	10.31	10.35	10.39	10.43	10.47	10.51	10.55	10.59	260
270	10.63	10.67	10.71	10.75	10.79	10.83	10.87	10.91	10.94	10.98	270
280	11.02	11.06	11.10	11.14	11.18	11.22	11.26	11.30	11.34	11.38	280
290	11.42	11.46	11.50	11.54	11.57	11.61	11.65	11.69	11.73	11.77	290

300	11.81	11.85	11.89	11.93	11.97	12.01	12.05	12.09	12.13	12.17	300
310	12.20	12.24	12.28	12.32	12.36	12.40	12.44	12.48	12.52	12.56	310
320	12.60	12.64	12.68	12.72	12.76	12.80	12.83	12.87	12.91	12.95	320
330	12.99	13.03	13.07	13.11	13.15	13.19	13.23	13.27	13.31	13.35	330
340	13.39	13.43	13.46	13.50	13.54	13.58	13.62	13.66	13.70	13.74	340
350	13.78	13.82	13.86	13.90	13.94	13.98	14.02	14.06	14.09	14.13	350
360	14.17	14.21	14.25	14.29	14.33	14.37	14.41	14.45	14.49	14.53	360
370	14.57	14.61	14.65	14.69	14.72	14.76	14.80	14.84	14.88	14.92	370
380	14.96	15.00	15.04	15.08	15.12	15.16	15.20	15.24	15.28	15.31	380
390	15.35	15.39	15.43	15.47	15.51	15.55	15.59	15.63	15.67	15.71	390
400	15.75	15.79	15.83	15.87	15.91	15.94	15.98	16.02	16.06	16.10	400
410	16.14	16.18	16.22	16.26	16.30	16.34	16.38	16.42	16.46	16.50	410
420	16.54	16.57	16.61	16.65	16.69	16.73	16.77	16.81	16.85	16.89	420
430	16.93	16.97	17.01	17.05	17.09	17.13	17.17	17.20	17.24	17.28	430
440	17.32	17.36	17.40	17.44	17.48	17.52	17.56	17.60	17.64	17.68	440
450	17.72	17.76	17.80	17.83	17.87	17.91	17.95	17.99	18.03	18.07	450

(continued)

(continued)

Conversion Table—Length
Millimeters to Inches (1 mm = 0.0394 in.)

Millimeters	0	1	2	3	4	5	6	7	8	9	Millimeters
460	18.11	18.15	18.19	18.23	18.27	18.31	18.35	18.39	18.43	18.46	460
470	18.50	18.54	18.58	18.62	18.66	18.70	18.74	18.78	18.82	18.86	470
480	18.90	18.94	18.98	19.02	19.06	19.09	19.13	19.17	19.21	19.25	480
490	19.29	19.33	19.37	19.41	19.45	19.49	19.53	19.57	19.61	19.65	490
500	19.69	19.72	19.76	19.80	19.84	19.88	19.92	19.96	20.00	20.04	500
510	20.08	20.12	20.16	20.20	20.24	20.28	20.31	20.35	20.39	20.43	510
520	20.47	20.51	20.55	20.59	20.63	20.67	20.71	20.75	20.79	20.83	520
530	20.87	20.91	20.94	20.98	21.02	21.06	21.10	21.14	21.18	21.22	530
540	21.26	21.30	21.34	21.38	21.42	21.46	21.50	21.54	21.58	21.61	540
550	21.65	21.69	21.73	21.77	21.81	21.85	21.89	21.93	21.97	22.01	550
560	22.05	22.09	22.13	22.17	22.20	22.24	22.28	22.32	22.36	22.40	560
570	22.44	22.48	22.52	22.56	22.60	22.64	22.68	22.72	22.76	22.80	570
580	22.83	22.87	22.91	22.95	22.99	23.03	23.07	23.11	23.15	23.19	580
590	23.23	23.27	23.31	23.35	23.39	23.43	23.46	23.50	23.54	23.58	590

600	23.62	23.66	23.70	23.74	23.78	23.82	23.86	23.90	23.94	23.98	600
610	24.02	24.06	24.09	24.13	24.17	24.21	24.25	24.29	24.33	24.37	610
620	24.41	24.45	24.49	24.53	24.57	24.61	24.65	24.68	24.72	24.76	620
630	24.80	24.84	24.88	24.92	24.96	25.00	25.04	25.08	25.12	25.16	630
640	25.20	25.24	25.28	25.31	25.35	25.39	25.43	25.47	25.51	25.55	640
650	25.59	25.63	25.67	25.71	25.75	25.79	25.83	25.87	25.91	25.94	650
660	25.98	26.02	26.06	26.10	26.14	26.18	26.22	26.26	26.30	26.34	660
670	26.38	26.42	26.46	26.50	26.54	26.57	26.61	26.65	26.69	26.73	670
680	26.77	26.81	26.85	26.89	26.93	26.97	27.01	27.05	27.09	27.13	680
690	27.17	27.20	27.24	27.28	27.32	27.36	27.40	27.44	27.48	27.52	690
700	27.56	27.60	27.64	27.68	27.72	27.76	27.80	27.83	27.87	27.91	700
710	27.95	27.99	28.03	28.07	28.11	28.15	28.19	28.23	28.27	28.31	710
720	28.35	28.39	28.43	28.46	28.50	28.54	28.58	28.62	28.66	28.70	720
730	28.74	28.78	28.82	28.86	28.90	28.94	28.98	29.02	29.06	29.09	730
740	29.13	29.17	29.21	29.25	29.29	29.33	29.37	29.41	29.45	29.49	740
750	29.53	29.57	29.61	29.65	29.68	29.72	29.76	29.80	29.84	29.88	750

(continued)

(continued)

Millimeters	0	1	2	3	4	5	6	7	8	9	Millimeters
760	29.92	29.96	30.00	30.04	30.08	30.12	30.16	30.20	30.24	30.28	760
770	30.31	30.35	30.39	30.43	30.47	30.51	30.55	30.59	30.63	30.67	770
780	30.71	30.75	30.79	30.83	30.87	30.91	30.94	30.98	31.02	31.06	780
790	31.10	31.14	31.18	31.22	31.26	31.30	31.34	31.38	31.42	31.46	790
800	31.50	31.54	31.57	31.61	31.65	31.69	31.73	31.77	31.81	31.85	800
810	31.89	31.93	31.97	32.01	32.05	32.09	32.13	32.17	32.20	32.24	810
820	32.28	32.32	32.36	32.40	32.44	32.48	32.52	32.56	32.60	32.64	820
830	32.68	32.72	32.76	32.80	32.83	32.87	32.91	32.95	32.99	33.03	830
840	33.07	33.11	33.15	33.19	33.23	33.27	33.31	33.35	33.39	33.43	840
850	33.46	33.50	33.54	33.58	33.62	33.66	33.70	33.74	33.78	33.82	850
860	33.86	33.90	33.94	33.98	34.02	34.06	34.09	34.13	34.17	34.21	860
870	34.25	34.29	34.33	34.37	34.41	34.45	34.49	34.53	34.57	34.61	870
880	34.65	34.68	34.72	34.76	34.80	34.84	34.88	34.92	34.96	35.00	880
890	35.04	35.08	35.12	35.16	35.20	35.24	35.28	35.31	35.35	35.39	890

Conversion Table—Length
Millimeters to Inches (1 mm = 0.0394 in.)

Square Feet to Square Meters 1 sq. ft. = 0.0929034 sq. m										
Square Feet	0	1	2	3	4	5	6	7	8	9
0	0.000	0.093	0.186	0.279	0.372	0.465	0.557	0.650	0.743	0.836
10	0.929	1.022	1.115	1.208	1.301	1.394	1.486	1.579	1.672	1.765
20	1.858	1.951	2.044	2.137	2.230	2.323	2.415	2.508	2.601	2.694
30	2.787	2.880	2.973	3.066	3.159	3.252	3.345	3.437	3.530	3.623
40	3.716	3.809	3.902	3.995	4.088	4.181	4.274	4.366	4.459	4.552
50	4.645	4.738	4.831	4.924	5.017	5.110	5.203	5.295	5.388	5.481
60	5.574	5.667	5.760	5.853	5.946	6.039	6.132	6.225	6.317	6.410
70	6.503	6.596	6.689	6.782	6.875	6.968	7.061	7.154	7.246	7.339
80	7.432	7.525	7.618	7.711	7.804	7.897	7.990	8.083	8.175	8.268
90	8.361	8.454	8.547	8.640	8.733	8.826	8.919	9.012	9.105	9.197

Square Meters to Square Feet 1 sq. m = 10.76387 sq. ft.										
Square Meters	0	1	2	3	4	5	6	7	8	9
0	0.00	10.76	21.53	32.29	43.06	53.82	64.58	75.35	86.11	96.87
10	107.64	118.40	129.17	139.93	150.69	161.46	172.22	182.99	193.75	204.51
20	215.28	226.04	236.81	247.57	258.33	269.10	279.86	290.62	301.39	312.15
30	322.92	333.68	344.44	355.21	365.97	376.74	387.50	398.26	409.03	419.79
40	430.56	441.32	452.08	462.85	473.61	484.37	495.14	505.90	516.67	527.43
50	538.19	548.96	559.72	570.49	581.25	592.01	602.78	613.54	624.30	635.07
60	645.83	656.60	667.36	678.12	688.89	699.65	710.42	721.18	731.94	742.71
70	753.47	764.23	775.00	785.76	796.53	807.29	818.05	828.82	839.58	850.35
80	861.11	871.87	882.64	893.40	904.17	914.93	925.69	936.46	947.22	957.98
90	968.75	979.51	990.28	1001.04	1011.80	1022.57	1033.33	1044.10	1054.86	1065.62

Conversion Table—Weights Pounds to Kilograms (1 lb = 0.4536 kg)										
Pounds	0	1	2	3	4	5	6	7	8	9
0	0.00	0.45	0.91	1.36	1.81	2.27	2.72	3.18	3.63	4.08
10	4.54	4.99	5.44	5.90	6.35	6.80	7.26	7.71	8.16	8.62
20	9.07	9.53	9.98	10.43	10.89	11.34	11.79	12.25	12.70	13.15
30	13.61	14.06	14.52	14.97	15.42	15.88	16.33	16.78	17.24	17.69
40	18.14	18.60	19.05	19.50	19.96	20.41	20.87	21.32	21.77	22.23
50	22.68	23.13	23.59	24.04	24.49	24.95	25.40	25.86	26.31	26.76
60	27.22	27.67	28.12	28.58	29.03	29.48	29.94	30.39	30.84	31.30
70	31.75	32.21	32.66	33.11	33.57	34.02	34.47	34.93	35.38	35.83
80	36.29	36.74	37.20	37.65	38.10	38.56	39.01	39.46	39.92	40.37
90	40.82	41.28	41.73	42.18	42.64	43.09	43.55	44.00	44.45	44.91

Kilograms to Pounds (1 kg = 2.2046 lbs)										
Kilograms	0	1	2	3	4	5	6	7	8	9
0	0.00	2.20	4.41	6.61	8.82	11.02	13.23	15.43	17.64	19.34
10	22.05	24.25	26.46	28.66	30.86	33.07	35.27	37.48	39.68	41.89
20	44.09	46.30	48.50	50.71	52.91	55.12	57.32	59.52	61.73	63.93
30	66.14	68.34	70.55	72.75	74.96	77.16	79.37	81.57	83.77	85.98
40	88.18	90.39	92.59	94.80	97.00	99.21	101.41	103.62	105.82	108.03
50	110.23	112.43	114.64	116.84	119.05	121.25	123.46	125.66	127.87	130.07
60	132.28	134.48	136.69	138.89	141.09	143.30	145.50	147.71	149.91	152.12
70	154.32	156.53	158.73	160.94	163.14	165.35	167.55	169.75	171.96	174.16
80	176.37	178.57	180.78	182.98	185.19	187.39	189.60	191.80	194.00	196.21
90	198.41	200.62	202.82	205.03	207.23	209.44	211.64	213.85	216.05	218.26

U.S. Gallons to Liters 1 U.S. gal = 3.785329 L										
Gallons	0	1	2	3	4	5	6	7	8	9
0	0	3.79	7.57	11.36	15.14	18.93	22.71	26.50	30.28	34.07
10	37.85	41.64	45.42	49.21	52.99	56.78	60.57	64.35	68.14	71.92
20	75.71	79.49	13.28	87.01	90.85	94.63	98.42	102.20	105.99	109.77
30	113.56	117.35	121.13	124.92	128.70	132.49	136.27	140.06	143.84	147.63
40	151.41	155.20	158.98	162.77	166.55	170.34	174.13	177.91	181.70	185.48
50	189.27	193.05	196.84	200.62	204.41	208.19	211.98	215.76	219.55	223.33
60	227.12	230.91	234.69	238.48	242.26	246.05	249.83	253.62	257.40	261.19
70	264.97	268.76	272.54	276.33	280.11	283.90	287.69	291.47	295.26	299.04
80	302.83	306.61	310.40	314.18	317.97	321.75	325.54	329.32	333.11	336.89
90	340.68	344.46	348.25	352.04	355.82	359.60	363.39	367.18	370.96	374.75

Liter to U.S. Gallon 1 L = 0.264168 U.S. gal										
Liters	0	1	2	3	4	5	6	7	8	9
0	0	0.26	0.53	0.79	1.06	1.32	1.59	1.85	2.11	2.38
10	2.64	2.91	3.17	3.43	3.70	3.96	4.23	4.49	4.76	5.02
20	5.28	5.55	5.81	6.08	6.34	6.60	6.87	7.13	6.60	7.66
30	7.93	8.19	8.45	8.72	8.98	9.25	9.51	9.77	10.04	10.30
40	10.57	10.83	11.10	11.36	11.62	11.89	12.15	12.42	12.68	12.94
50	13.21	13.47	13.74	14.00	14.27	14.53	14.79	15.06	15.32	15.59
60	15.85	16.11	16.38	16.64	16.91	17.17	17.44	17.70	17.96	18.23
70	18.49	18.76	19.02	19.28	19.55	19.81	20.08	20.34	20.61	20.87
80	21.13	21.40	21.66	21.93	22.19	22.45	22.72	22.98	23.25	23.51
90	23.78	24.04	24.30	24.57	24.83	25.10	25.36	25.62	25.89	26.15

Conversion Factors (For conversion factors meeting the standards of the SI metric system, refer to ASTM E380-72)		
Multiply	**by**	**To Obtain**
Centimeters	3.28083×10^{-2}	Feet
Centimeters	0.3937	Inches
Cubic centimeters	6.102×10^{-2}	Cubic inches
Cubic feet	2.8317×10^{-2}	Cubic meters
Cubic feet	6.22905	Gallons, British Imperial
Cubic feet	28.3170	Liters
Cubic inches	16.38716	Cubic centimeters
Cubic meters	35.3145	Cubic feet
Cubic meters	1.30794	Cubic yards
Cubic yards	0.764559	Cubic meters
Degrees angular	0.0174533	Radians
Foot-pounds	0.13826	Kilogram-meters
Feet	30.4801	Centimeters
Gallons, British Imperial	0.160538	Cubic feet
Gallons, British Imperial	1.20091	Gallons, U.S.
Gallons, British Imperial	4.54596	Liters
Gallons, U.S.	0.832702	Gallons, British Imperial
Gallons, U.S.	0.13368	Cubic feet
Gallons, U.S.	3.78543	Liters
Grams, metric	2.20462×10^{-3}	Pounds, avoirdupois
Horsepower, metric	0.98632	Horsepower, U.S.
Horsepower, U.S.	1.01387	Horsepower, metric
Inches	2.54001	Centimeters
Kilograms	2.20462	Pounds
Kilograms per square centimeter	14.2234	Pounds per square inch
Kilometers	0.62137	Miles, statute
Liters	0.26417	Gallons, U.S.
Meters	3.28083	Feet
Meters	39.37	Inches
Meters	1.09361	Yards
Miles, statute	1.60935	Kilometer
Millimeters	3.28083×10^{-3}	Feet
Millimeters	3.937×10^{-2}	Inches
Pounds avoirdupois	0.453592	Kilograms
Pounds per square foot	4.88241	Kilograms per square meter
Pounds per square inch	7.031×10^{-2}	Kilograms per square centimeter
Radians	57.29578	Degrees angular
Square centimeters	0.1550	Square inches

(*continued*)

(continued)

Conversion Factors (For conversion factors meeting the standards of the SI metric system, refer to ASTM E380-72)

Multiply	by	To Obtain
Square inches	6.45163	Square centimeters
Square meters	1.19599	Square yards
Square miles	2.590	Square kilometers
Square yards	0.83613	Square meters
Tons, long	1016.05	Kilograms
Tons, long	2240.00	Pounds
Tons, metric	2204.62	Pounds
Tons, metric	0.98421	Tons, long
Tons, metric	1.10231	Tons, short
Tons, short	0.892857	Tons, long
Tons, short	0.907185	Tons, metric
Yards	0.914402	Meters

Index

Printed and bound by CPI Group (UK) Ltd, Croydon, CR0 4YY

18/10/2024

01776270-0007